Multi-Source Industrial Solid Waste Synergistically Utilized in Manganese Slag-Based Grouting Materials

Zhonghu Wu • Chi Zeng • Yuan Liu
Zhaofeng Li • Shuailong Jia

Multi-Source Industrial Solid Waste Synergistically Utilized in Manganese Slag-Based Grouting Materials

Springer

Zhonghu Wu
School of Civil Engineering
Guizhou University
Guiyang, Guizhou, China

Yuan Liu
Guizhou Tianwei Building Mat's Tech. Co.
Guiyang, Guizhou, China

Shuailong Jia
College of Civil Engineering
Hebei University of Engineering
Handan, Hubei, China

Chi Zeng
Guizhou Tianwei Building Mat's Tech. Co.
Guiyang, Guizhou, China

Zhaofeng Li
Qilu Transportation College
Shandong University
Jinan, Shandong, China

ISBN 978-981-92-3020-4 ISBN 978-981-92-3021-1 (eBook)
https://doi.org/10.1007/978-981-92-3021-1

This Springer imprint is published by the registered company Springer Nature Singapore Pte Ltd.
The registered company address is: 152 Beach Road, #21-01/04 Gateway East, Singapore 189721, Singapore

Preface

As the world's major producer and consumer of electrolytic manganese, China has accumulated more than 100 million tons of electrolytic manganese residue (EMR), with an annual increase of 13 million tons. Large-scale stockpiling of EMR has posed severe ecological and environmental risks. Featuring a complex chemical composition and difficult disposal, EMR is characterized by low utilization efficiency and low added value in conventional resource utilization routes, which has become a critical bottleneck restricting the green development of the manganese industry. Meanwhile, huge amounts of multi-source industrial solid wastes such as blast furnace slag, steel slag, fly ash, and coal slag are produced, while their synergistic utilization level remains low. Therefore, new technologies for efficient, high-value, and large-scale utilization are urgently needed.

Grouting is a core technique for geotechnical reinforcement, seepage prevention, and water blocking. Traditional cement-based grouting materials suffer from inherent drawbacks, including high energy consumption, large carbon emissions, low early strength, and insufficient durability, which can hardly meet the requirements of the "dual-carbon" strategy and green engineering. To address the national demands for solid waste reduction and low-carbon construction, the research on the theory and key technologies of preparing manganese residue-based grouting materials through synergistic utilization of multi-source industrial solid wastes is of great scientific significance and engineering application potential.

This book systematically elaborates the mechanisms of thermal activation, ammonia and sulfur removal, and cementitious activity excitation of electrolytic manganese residue. It establishes an alkali-activated grouting material system composed of EMR synergistically utilized with blast furnace slag, fly ash, steel slag, and coal slag and reveals the workability, mechanical properties, microstructure, and hydration mechanism of three types of grouts, namely EFGM, ESGM, and ECGM. This book also clarifies the regulation rules of admixtures, the durability against ion erosion, and the solidification mechanisms of heavy metals and alkaline components and establishes a theoretical model for diffusion-coupling-reinforcement of grouting materials. With rigorous theories and sufficient data, this book forms a complete set of design, preparation, evaluation, and application methods for multi-source solid waste-based grouting materials, which can provide theoretical support and technical guidance for large-scale consumption of industrial solid wastes and the development of green grouting engineering.

This book summarizes years of research findings and engineering practice of the author team. During the writing, a comprehensive review and citation of authoritative literatures and cutting-edge progress worldwide have been made to pursue systematic integrity, substantial content, and advanced technology. Given the rapid development in the fields of industrial solid waste recycling and green grouting materials, as well as the limitation of the authors' knowledge, omissions and imperfections may remain in the book. Criticisms and suggestions from experts, peers, and readers are sincerely appreciated.

Guiyang, Guizhou, China Zhonghu Wu
Guiyang, Guizhou, China Chi Zeng
Guiyang, Guizhou, China Yuan Liu
Jinan, Shandong, China Zhaofeng Li
Handan, Hubei, China Shuailong Jia
January 2023

Acknowledgments

This research was supported by the National Natural Science Foundation of China (Project No.: 52464004) and the Guizhou Provincial Science and Technology Support Program (Project Nos.: Qiankehe Support [2025] General 080, Qiankehe Support [2025] General 078, Qiankehe Support [2023] General 137, and Qiankehe Support [2023] General 136).

We extend our sincere gratitude to the School of Civil Engineering at Guizhou University for providing us with an excellent working environment and research conditions. Thanks to the inspiring academic discussion atmosphere and the open, inclusive research ecosystem fostered by the research team composed of outstanding master's and doctoral students, our research work has been carried out smoothly. Special thanks are extended to Dr. Anfa Long and Dr. Kunpeng Zhang for their strong support in the finalization of this book. We also extend our gratitude to the relevant institutions for authorizing the use of certain chart materials.

Finally, we extend our sincere gratitude to the publisher Springer Nature Singapore Pte Ltd. and thank the editor Wayne Hu for his invaluable support throughout the writing of this book.

Competing Interests The authors have no competing interests to declare that are relevant to the content of this manuscript.

Contents

1 **Introduction** ... 1
  1.1 Research Background and Significance ... 1
  1.2 Current Research Status at Home and Abroad ... 3
    1.2.1 Current Research Status of Resource Utilization of Electrolytic Manganese Slag ... 3
    1.2.2 Research Status of Synergistic Preparation of Cementitious Materials from Multiple Sources of Industrial Solid Wastes ... 6
    1.2.3 Research Status of Industrial Solid Waste Grouting Materials ... 8
    1.2.4 The Shortcomings of the Existing Research ... 9
  1.3 Research Content and Technical Approach ... 10
    1.3.1 Research Content ... 10
    1.3.2 Research Route ... 11
  References ... 13

Part I Study on Pretreatment and Modification of Electrolytic Manganese Slag

2 **Research on Desulfurization Denitrification and Activation of Gelling Activity of Electrolytic Manganese Slag** ... 19
  2.1 Introduction ... 19
  2.2 Physical and Chemical Properties of Electrolytic Manganese Slag ... 19
    2.2.1 Physical Properties ... 20
    2.2.2 Chemical Composition ... 20
    2.2.3 Mineral Phase Composition ... 21

2.3 Research on Desulfurization, Denitrification, and Gelation Activity Activation of Electrolytic Manganese Slag 21
2.3.1 Manganese Slag Thermal Activation Treatment 21
2.3.2 The Influence of Thermal Activation Treatment on the Activity of Manganese Slag 24
2.3.3 The Influence of Alkaline Environment on the Cementing Activity of Manganese Slag. 27
2.4 Chapter Summary 29
References 29

Part II Basic Performance Evaluation of Manganese Slag-Based Grouting

3 Performance Evaluation of Electrolytic Manganese Residue-Based Grouting Materials 33
3.1 Introduction 33
3.2 Preparation of EFGM: An EMR-Based Grouting Material Incorporating GGBS and Fly Ash 34
3.2.1 Fresh Properties of the Grout 34
3.2.2 Strength Characteristics of the Grouting Material 36
3.2.3 Microstructural Characteristics of the Grouting Material ... 37
3.3 Preparation of ESGM: An EMR-Based Grouting Material Incorporating GGBS and Steel Slag 47
3.3.1 Fresh Properties of the Grout 47
3.3.2 Strength Characteristics of the Grouting Material 51
3.3.3 Microstructural Characteristics of the Grouting Material ... 56
3.4 Preparation of ECGM: An EMR-Based Grouting Material Incorporating GGBS and Coal Slag 66
3.4.1 Fresh Properties of the Grout 66
3.4.2 Strength Characteristics of the Grouting Material 70
3.4.3 Microstructural Characteristics of the Grouting Material 73
3.5 Chapter Summary 82
References 85

Part III Regulation of Admixtures on Performance of Grouting Materials

4 Effects of Admixtures on the Performance of EMR-Based Grouting Materials 91
4.1 Introduction 91
4.2 Effect of Admixtures on the Performance of EFGM 92
4.2.1 Effect of Admixtures on the Fresh Properties of EFGM Grout 92
4.2.2 Effect of Admixtures on the Strength Characteristics of EFGM 95
4.2.3 Effect of Admixtures on the Microstructural Characteristics of EFGM 98

4.3 Effect of Admixtures on the Performance of ESGM . . . 113
4.3.1 Effect of Admixtures on the Fresh Properties of EFGM Grout . . . 113
4.3.2 Effect of Admixtures on the Strength Characteristics of ESGM . . . 117
4.3.3 Effect of Admixtures on the Microstructural Characteristics of ESGM . . . 119
4.4 Effect of Admixtures on the Performance of ECGM . . . 125
4.4.1 Effect of Admixtures on the Fresh Properties of ECGM Grout . . . 125
4.4.2 Effect of Admixtures on the Strength Characteristics of ECGM . . . 129
4.4.3 Effect of Admixtures on the Microstructural Characteristics of ECGM . . . 130
4.5 Chapter Summary . . . 138
References . . . 139

Part IV Long-Term Performance and Prediction of Manganese Slag-Based Grouting Materials

5 Study on the Durability and Environmental Compatibility of Manganese Slag-Based Grouting Materials . . . 145
5.1 Introduction . . . 145
5.2 Effect of Ion Erosion on Strength Characteristics of Manganese Slag-Based Grouting Materials . . . 146
5.2.1 Effect of Ion Erosion on Strength Characteristics of EFGM . . . 146
5.2.2 Effect of Ion Erosion on the Strength Characteristics of ESGM . . . 149
5.2.3 Effect of Ion Erosion on ECGM Intensity Characteristics . . . 151
5.3 Effect of Ion Erosion on the Microstructural Characteristics of Manganese Slag-Based Grouting Materials . . . 153
5.3.1 Effect of Ion Erosion on the Microstructural Characteristics of EFGM . . . 153
5.3.2 Effect of Ion Erosion on the Microstructural Characteristics of ESGM . . . 163
5.3.3 Effect of Ion Erosion on the Microstructural Characteristics of ECGM . . . 171
5.4 Study on Heavy Metal Solidification Mechanisms of Manganese Slag-Based Grouting Materials . . . 177
5.4.1 Study on the Heavy Metal Solidification Mechanism of EFGM . . . 177
5.4.2 Study on the Heavy Metal Immobilization Mechanism of ESGM . . . 178
5.4.3 Study on the Heavy Metal Immobilization Mechanism of ECGM . . . 178

5.5 Study on the Curing Mechanism of the Alkali Component in Manganese Slag-Based Grouting Materials. 180
5.5.1 Study on Leaching Patterns of EFGM Alkaline Components 180
5.5.2 Study on Leaching Patterns of Alkaline Components in ESGM 182
5.5.3 Study on Leaching Patterns of ECGM Alkaline Components 183
5.5.4 SEM-EDS Analysis of Leaching Alkali Components 185
5.6 Chapter Summary 186
References 187

6 Performance Prediction Model for Manganese Slag-Based Grouting Materials 189
6.1 Introduction. 189
6.2 Study on the Diffusion Mechanism of Manganese Slag-Based Grouting Materials in Rock and Soil Masses. 190
6.2.1 Basic Assumptions. 190
6.2.2 Physical Field Control Equations and Geometric Construction. 191
6.2.3 Influence of Rock Mass Type on Diffusion Patterns 193
6.3 Study on the Coupling Mechanism Between Manganese Slag-Based Grouting Materials and Geomaterials. 197
6.3.1 Physical Field Control Equations. 197
6.3.2 Geometry and Mesh Models 199
6.3.3 Revealing the Coupling Mechanism Between Grouting Materials and Geomaterials 200
6.4 Study on the Reinforcement Mechanism of Manganese Slag-Based Grouting Materials in Water-Rich Karst Regions 201
6.4.1 Physical Field Control Equations. 202
6.4.2 Geometry and Mesh Models 204
6.4.3 Mechanism of Reinforcement in Water-Saturated Karst . . . 205
6.5 Subsections of this Chapter 208
References 209

7 Conclusions 211

Chapter 1
Introduction

1.1 Research Background and Significance

China holds a leading position in the global production of electrolytic manganese. Approximately 80% of the world's electrolytic manganese output comes from China. Currently, the amount of manganese slag accumulated in China has exceeded 100 million tons, and an additional 13 million tons are added each year, accounting for over 98% of the global output [1, 2]. The main distribution areas are Guangxi, Hunan, Guizhou, Yunnan, Chongqing, and Liaoning, which are rich in manganese resources. It is estimated that for every ton of manganese metal produced, approximately 7–10 tons of waste residue will be generated [3–5]. In China, most electrolytic manganese producers adopt the wet storage method to handle electrolytic manganese residues [6]. This open-air storage method not only consumes a large amount of land resources but also the harmful substances in the electrolytic manganese residues are highly likely to seep into the surrounding water bodies and soil, posing a serious threat to the local ecological environment [7, 8]. Therefore, it is particularly urgent to develop economic, efficient, and environmental-friendly technologies for the harmless and resourceful treatment of electrolytic manganese slag, to accelerate the processing of the accumulated electrolytic manganese slag, promote the sustainable development of China's electrolytic manganese slag industry and related downstream industries, and thereby indirectly contribute to the growth of the regional economy [9]. Songtao County in Guizhou Province is rich in manganese resources and is part of the "Manganese Triangle" core area. In July 2021, the National Development and Reform Commission, the Ministry of Industry and Information Technology, the Ministry of Natural Resources, and the Ministry of Ecology and Environment jointly issued the "Notice on Strengthening Manganese Pollution Control and Promoting the Adjustment of the Manganese Industry Structure." In November of the same year, the "Opinions on Deepening the Battle Against Pollution Prevention and Control" issued by the Central Committee of the

Z. Wu et al., *Multi-Source Industrial Solid Waste Synergistically Utilized in Manganese Slag-Based Grouting Materials*,
https://doi.org/10.1007/978-981-92-3021-1_1

Communist Party of China and the State Council clearly proposed to comprehensively improve the pollution control in the "Manganese Triangle" area in the Wuling Mountains region at the junction of Chongqing, Hunan and Guizhou provinces. On March 27, 2022, the Ministry of Ecology and Environment released the "Technical Specifications for Pollution Control of Manganese Sludge," which is the first national ecological environment standard in China specifically formulated for the pollution control technology of manganese sludge. This specification details the pollution control technical requirements for manganese sludge during collection, storage, transportation, pre-treatment, utilization, filling, backfilling, and landfilling, as well as the relevant requirements for monitoring and environmental management. Manganese slag belongs to the second category of general industrial solid waste. Its moisture content ranges from 20% to 30%. It contains a large amount of soluble sulfates such as manganese sulfate and ammonium sulfate, as well as heavy metals such as zinc, copper, nickel, and cobalt. The composition is complex and the treatment is very difficult [10, 11]. With the rapid development of industries such as metallurgy, coal mining, and forging, a large amount of industrial solid waste such as slag, steel slag, and fly ash has been produced. Due to the failure to handle and utilize them in a timely and effective manner, they have been discarded and piled up disorderly. Therefore, the goal proposed by the Guizhou Provincial People's Government in the "Guizhou Province's 14th Five-Year Plan for National Economic and Social Development and Long-Term Goals until 2035"—that "by 2025, the comprehensive utilization rate of industrial solid waste will reach 70%"—allows for the harmless disposal and comprehensive utilization of the filter residue from electrolytic manganese production, which can absorb a large amount of electrolytic manganese waste and turn it into a resource, aligning with national policy directions and the development plans of Guizhou Province [12].

Industrial solid waste mainly refers to the solid waste generated during industrial manufacturing processes, including electrolytic manganese slag, red mud, coal slag, fly ash, steel slag, etc. [13]. The main characteristics of industrial solid waste are as follows: (1) Under the influence of the external environment, it is very likely to cause continuous pollution to the surrounding environment, and this pollution has a long-term and persistent nature; (2) Industrial solid waste generally has strong concealment and is difficult to be discovered, and its duration is long [14]. The requirements of ecological environment protection and the circular development of the national economy call for the resource utilization of industrial waste such as electrolytic manganese slag and coal slag, reducing the environmental pollution caused by industrial solid waste materials, and increasing the utilization rate of solid waste in the field of building materials, which is in line with the national strategic requirements.

Grouting, as a key method for enhancing the physical and mechanical properties of rock and soil bodies and achieving the goal of anti-seepage reinforcement of surrounding rocks, has been widely applied in the fields of geological and tunnel engineering disaster prevention [15]. Grouting materials, as the main means for dealing with poor geological bodies, play a crucial role in grouting projects. At present, silicate cement-based grouting materials are the most widely used. However,

Fig. 1.1 Manganese slag storage area

traditional cement-based grouting materials have high energy consumption and high carbon emissions, as well as poor early strength growth rate and other engineering performance defects that are becoming increasingly prominent. In response to the "dual carbon" policy, this research aims to provide a new idea for a low-energy consumption and low-cost new type of grouting material. That is, based on a thorough understanding of the reaction mechanism of industrial waste materials' gelling materials under alkaline stimulation, this study developed alkaline-stimulated industrial waste-based grouting slurry. Through laboratory tests, the engineering properties of industrial waste-based grouting materials were evaluated to demonstrate the feasibility of the alkaline-stimulated industrial waste-based grouting material technology and the superiority of the slurry performance. In addition, the flow pattern, rheological parameters, and time-dependent characteristics of the industrial waste-based slurry were tested, with the aim of laying a solid theoretical foundation for the future engineering application of industrial waste-based slurry (Fig. 1.1).

1.2 Current Research Status at Home and Abroad

1.2.1 Current Research Status of Resource Utilization of Electrolytic Manganese Slag

1.2.1.1 Cement and Concrete

The main minerals of electrolytic manganese slag exhibit excellent crystallization properties. XRD analysis indicates that its main phase composition is gypsum dimer and quartz. Among them, the proportion of the hydrated gypsum phase is as high as 25.9%, contributing 60%–70% of the total sulfate content of the system. Based on this characteristic, researchers applied the strategy of using high-sulfate solid waste as a substitute to the super-sulfate cement system. For instance, Tang et al. [16] utilized the sulfate components in electrolytic manganese slag, combined with carbonized slag and granulated blast furnace slag, to prepare a full solid waste electrolytic manganese slag super-sulfate cement. The optimal composition was 4% carbonized slag, 40% electrolytic manganese slag, and 56% granulated blast furnace slag. At 28 days, the compressive strength reached 38.34 MPa, meeting the requirements of 32.5 grade cement.

The electrolytic manganese slag contains a certain amount of magnesium salts and calcium sulfate. The electrolytic manganese slag can be calcined to produce sulfur-aluminate cement clinker, and then sulfur-aluminate cement can be prepared [17]. Hou et al. [18] used electrolytic manganese slag, limestone, kaolin, etc. and heated them at 1100–1270 °C. After adding a certain amount of gypsum, they prepared a class of sulfur aluminosilicate cement with a compressive strength of up to 65 MPa after 56 days. Reje [19] et al. used electrolytic manganese slag and Al_2O_3 to prepare high-temperature sulfur-aluminum phosphate cement by calcination at 1200 °C. The 3-day compressive strength of this cement was approximately 50 MPa. Zhao Shizhen et al. [20] used electrolytic manganese slag, magnesium slag, and lime to prepare sulfur-aluminum phosphate cement clinker at 1260 °C, with 28-day flexural and compressive strengths reaching 5.1 MPa and 31.2 MPa respectively. He et al. [21] used electrolytic manganese slag, barium slag, limestone, and bauxite as raw materials and calcined at 1300 °C to prepare cement, with a 28-day compressive strength as high as 60 MPa.

Unactivated electrolytic manganese slag has relatively low chemical reactivity and can be used as a concrete filler. Ye Fen et al. [22] used electrolytic manganese slag to prepare ceramic aggregates. The aggregates combined with cementitious materials more closely, and the flexural strength of the concrete specimens prepared with these aggregates was higher than that of the specimens prepared with natural sand and gravel. Xu et al. [23] found that adding electrolytic manganese slag into ultra-high performance concrete can increase the initial setting time and enhance the compactness of the hydration products. He et al. [24] utilized concrete containing a 10% substitution rate of silica-manganese slag, which achieved a compressive strength of 48.15 MPa. Moreover, the slump and salt-frost resistance of this concrete were also significantly improved. Tang et al. [25] used electrolytic manganese slag, slag, red mud and tailings, and sodium hydroxide as the alkali to prepare a cemented filling material. By adding 20% of electrolytic manganese slag, the compressive strength reached 11.04 MPa after 28 days. Yu Yunlin et al. [26] prepared a mine filling material without adding cement by electrolyzing manganese slag, phosphogypsum, fly ash and calcium hydroxide and then grinding them together. When the content of electrolytic manganese slag was 65%, the performance of the prepared samples, such as the water permeability rate and compressive strength, was effectively improved.

1.2.1.2 Wall Material

The main wall materials include bricks, blocks, and plates, which are widely used in construction projects. Electrolytic manganese slag is used to make bricks as shown in Fig. 1.2. Utilizing industrial solid waste as raw materials to produce wall materials can achieve the resource utilization of industrial solid waste and also reduce the

Fig. 1.2 Brick production from electrolytic manganese slag

production cost of wall materials. Luo et al. [27] used electrolytic manganese slag and fly ash to manufacture concrete blocks. By replacing part of the sand with electrolytic manganese slag, the volume of permeable voids could be significantly reduced, and the compressive strength of the concrete blocks under all curing ages could be enhanced. Zhou Youlian et al. [28] used shale, limestone and fly ash in combination with electrolytic manganese slag to prepare fired bricks. When the content of electrolytic manganese slag was 50% to 70%, the performance of the bricks was the best. Yang Hongyou et al. [29] used pre-treated electrolytic manganese slag and silica ore to prepare non-fired bricks. When the mass ratio of electrolytic manganese slag to silica ore was 4:1, the compressive strength of the bricks prepared was 11.25 MPa, and its performance met the requirements of relevant standards. Wang et al. [30] prepared permeable bricks using electrolytic manganese slag, blast furnace slag, cement, crushed stones, and additives. The content of electrolytic manganese slag in these bricks could reach up to 60%, and they were successfully used in pavement construction projects. You Xiaoyu et al. [31] used pre-treated electrolytic manganese slag and cement to prepare non-fired bricks that met the relevant standard requirements. The content of pre-treated electrolytic manganese slag in these non-fired bricks could reach 52%. Lan et al. [32] achieved the activation of silicon in electrolytic manganese slag through mechanical activation, promoted NH_4^+ removal, and prepared non-fired bricks with a compressive strength of 12 MPa.

1.2.1.3 Other Building Materials

Electrolytic manganese slag can also be used as raw materials for preparing roadbed materials, ceramics, and geopolymer and other construction materials, as shown in Fig. 1.3. Cai Mingying et al. [33] demonstrated through experiments combining different ratios of manganese slag and lime that it can significantly reduce the pollutant content of electrolytic manganese slag, and the moisture content can be completely controlled. This not only significantly improves the solidification effect of the roadbed but also achieves good social and economic benefits. He et al. [21] used EMR to reduce the utilization rate of cement-based materials and demonstrated that when the dosage of EMR was between 10% and 15%, the carbon emission index of the

Fig. 1.3 Roadbed made of electrolytic manganese slag

hydration product could be significantly reduced. Qian et al. [34] studied the direct utilization of electrolytic manganese slag for calcination at 1350 °C to prepare glass ceramics. Wang Gongxun et al. [35] prepared recycled ceramic tiles under the condition of a mass ratio of electrolytic manganese slag and waste ceramic grinding powder of 1:9, and investigated the firing process of the recycled ceramic tiles. Han et al. [36] conducted research and found that when the weight ratio of river sand to electrolytic manganese slag was 0.8, a geopolymer with a compressive strength of 96.30 MPa was prepared using phosphoric acid as the activator. MI et al. [37] used manganese slag, coal gangue, and fly ash as raw materials to produce geopolymer. The leaching concentrations of the main heavy metals manganese and ammonia nitrogen in the geopolymer were 0.123 mg/L and 0.538 mg/L respectively, both of which were lower than the national standards. Zhan et al. [38] prepared geopolymer using electrolytic manganese slag pre-treated with NaOH solution and calcined slag from waste incineration, and this method could also solidify the heavy metal ions in both raw materials.

Although certain results have been achieved in the field of construction, due to the high content of Mn^{2+}, NH_4^+, and SO_4^{2-} and the low activity of these ions in the electrolytic manganese slag, they cannot be directly utilized. This results in a relatively small usage in construction projects, preventing the large-scale application of electrolytic manganese slag and making it difficult to break through the technical bottleneck of manganese slag resource utilization. Therefore, there is still a lot of work to be done to select more economically reasonable methods to achieve efficient utilization of electrolytic manganese slag.

1.2.2 Research Status of Synergistic Preparation of Cementitious Materials from Multiple Sources of Industrial Solid Wastes

As society continues to develop and progress, the accumulation of industrial solid waste has become a difficult problem in terms of environmental pollution and treatment. Scholars are working to transform these wastes into useful resources, such as using industrial solid waste to manufacture cementitious materials, which is beneficial to environmental protection [39–42]. However, due to the generally low cementitious activity of industrial solid wastes, direct utilization is difficult. Therefore, in most cases, silicate cement is needed as a binder, or external agents are added or high-temperature treatment is carried out to enhance their activity, in order to effectively fix heavy metal ions [43, 44].

Taking the addition of cement or finished clinker as an example, He et al. [45] used a ternary mixed cementitious material composed of active electrolytic manganese slag, slag and cement as the main raw materials, and the compressive strength at 28 days exceeded 44 MPa. Ding et al. [46] prepared persulfate phosphogypsum cement by using slag, steel slag, phosphogypsum (PG) and clinker. The microporous structure of this cement is better than that of traditional silicate cement, and it has obvious advantages in corrosion resistance; Li et al. [47] produced phosphogypsum-based foam concrete using phosphogypsum, cement, and slag. Persulfate cement outperformed the cement control group in durability performance; the addition of fibers can improve the toughness of the cementitious material. Hua et al. [48] prepared cementitious materials using phosphogypsum, slag and an appropriate amount of fibers. It was found that adding mineral fibers, glass and polypropylene fibers significantly improved the flexural strength, impact energy and water resistance of the mixture.

Mixing different types of waste in an orderly manner seems to enhance the cementation activity of each type. Chen et al. [49] prepared solid waste cementitious materials using slag, steel slag, and improved semi-hydrated phosphogypsum. Compared with pure HPG cementitious materials, adding an appropriate amount of slag and steel slag increased the compressive strength and softening coefficient of the cementitious materials. This has guiding significance for the comprehensive utilization of HPG, slag, and steel slag. Feng et al. [50] prepared alkaline-activated cementitious materials using electrolytic manganese slag, steel slag, and blast furnace slag. The compressive strength of these materials increased initially with the increase in EMR and the concentration of the alkaline activator, but then decreased. The compressive strength after 28 days was 23 MPa. Zhou et al. [51] used aluminosilicate cement, mineral powder, desulfurized gypsum, and a certain amount of semi-hydrated desulfurized gypsum as the cementitious agents to prepare desulfurized gypsum-based composite cementitious materials. They confirmed that the added cement and mineral powder had a direct impact on the flexural strength and softening coefficient of the cementitious materials.

Currently, the research of many scholars mainly focuses on using cement as a binder and combining various industrial solid wastes [52, 53] to prepare materials with superior performance. For instance, some materials exhibit stronger corrosion resistance, while others achieve lower costs and lower carbon emissions. These achievements not only produce lighter building materials but also make positive contributions to the sustainable development of society. Thus, it can be seen that the theoretical feasibility of preparing cementitious materials by combining multiple industrial solid wastes with manganese slag is completely achievable.

1.2.3 Research Status of Industrial Solid Waste Grouting Materials

Grouting refers to the process of injecting grouting slurry into the ground layer or the weak rock and soil layer that needs reinforcement to prevent water leakage and surface subsidence [54]. Grouting has a good effect on reinforcing, preventing seepage in large dams, tunnels and foundation pits, and has been well applied in these fields [55]. Grouting materials are one of the most crucial elements in the grouting process. The quality and performance of grouting materials directly affect the grouting reinforcement and seepage prevention effects. However, the performance of grouting slurry is the most important aspect of grouting materials. The performance of grouting is related to whether the grouting materials can be completely injected into the rock and soil layer, and plays an important role in the quality and success or failure of the grouting project [56]. The research on grouting projects in China started relatively late, but after many years of development, a large number of achievements have been made. Currently, scholars at home and abroad have conducted a large number of experiments and analyses on grouting materials and have applied them to actual projects, achieving good results.

In China, industrial solid-waste-based grouting materials have achieved certain results. Dong Xiaojuan [57] used high-temperature blast furnace slag and coal gangue and other waste materials to prepare an environmental-friendly grouting material with good strength and anti-seepage performance and meeting the requirements of grouting setting time. Xiao Jie et al. [58] used three types of industrial solid waste: desulfurization fly ash from circulating fluidized bed boilers, steel slag micropowder, and desulfurization gypsum to prepare a multi-source solid waste collaborative grouting material. Through indoor experiments and systematic analysis of micro-characteristics, the influence of raw material ratios and water-cement ratios on the working performance and mechanical properties of grouting materials was studied. The research results can provide green and low-cost grouting materials for the treatment of collapsed areas beneath roadbeds. Wei Jing [59] used steel slag and desulfurized gypsum as modification materials, and used cement mortar as the geopolymer matrix to prepare a new type of environmentally friendly and low-cost grouting material.

Abroad, Tamburini et al. [60] used raw materials such as blast furnace slag and high-kaolin to prepare sodium silicate geopolymer grouting cementitious materials. The results showed that the newly prepared grouting cementitious materials had good fluidity and corrosion resistance, providing a reference for the development of grouting materials in harsh environments. Song et al. [61], based on the performance requirements of tunnel shield wall grouting materials, prepared a new type of two-component grouting cementitious material. At the same time, they tested the setting time, mechanical strength, and fluidity of the grout solution, and found that it had good engineering applicability and was effective for applying to tunnel shield wall grouting. Wang et al. [62] studied the influence of cement-gypsum grouting material on the segment structure during synchronous grouting for shield tunneling filling. By using random finite element numerical calculation, they derived the reduction factor for the mechanical properties of the cement-gypsum slurry reaction products after degradation. Within the design range, the elastic modulus reduction coefficient decreases as the amount of gypsum increases, and the reduction rate of the reduction coefficient increases as the amount of gypsum increases. Guo Chiwu et al. [63], based on the analysis of the chemical reaction mechanism of cement-silica gel, studied the effects of two admixtures, namely aluminum chloride solution and disodium hydrogen phosphate, on the solidification characteristics of the slurry. The results showed that both admixtures had a promoting effect on the compressive strength of the hardened paste, and the promoting effect of the aluminum chloride solution was more significant than that of disodium hydrogen phosphate. Zhang [64] et al. studied the rheological properties of rapid-setting grouting during the phase transition of cement-sodium silicate slurry to solid. They found that the gel reaction process of the cement (C)—silica gel (S) slurry could be divided into three stages: liquid phase, liquid–solid phase, and solid phase. The gel reaction speed of the cement-silica gel slurry during the reaction process was positively correlated with the volume ratio of C/S [65, 66], and negatively correlated with the ratio of water and cement mass.

1.2.4 The Shortcomings of the Existing Research

In recent years, grouting materials have received widespread attention and numerous studies have been conducted, especially in the development of cementitious materials using industrial waste. Remarkable achievements have been made, and some industrial waste has been extensively utilized in construction projects, such as slag, fly ash, and phosphogypsum. Although significant efforts have been made in the fields of grouting materials, EMR, and the resource utilization of solid waste materials, there are still many shortcomings:

1. In China, over 98% of manganese slag disposal adopts the method of storage in slag repositories. Although slag storage is the least costly method, the manganese ore in China has a low grade and low resource utilization rate, resulting in abundant EMR when producing manganese metal from the ore. Currently, the utilization rate of EMR in industrial materials is still at a low level, with the national EMR resource utilization rate being approximately 15%–20%. In the industrial materials sector (such as building materials, ceramics, etc.), the contribution rate of EMR is only 5%–8%. Traditional cement-based grouting materials have high CO_2 emissions and do not conform to the green construction concept. Therefore, promoting the comprehensive utilization and reduction of EMR is imperative, and it is necessary to research environmentally friendly resource utilization technologies.
2. The electrolytic manganese slag may contain some harmful components, such as soluble heavy metals (Mn, Zn, Cu, Pb, Cd, Cr) and high-risk environmental substances like ammonia nitrogen. By using lime, sodium hydroxide, potassium hydroxide, etc. as alkaline additives to treat the electrolytic manganese slag, the

removal of Mn^{2+} and ammonia nitrogen can be achieved. The chemical treatment technology can efficiently recycle and utilize the manganese slag resources with low cost, but it is only applicable in the laboratory and there are no application examples in industry. During the co-production of cementitious materials, the existence forms, migration and transformation laws of these harmful components, as well as their impacts on the performance and environmental safety of the cementitious materials, still need further research. If these harmful components cannot be effectively treated, it will limit the application of manganese slag-based cementitious materials in engineering.

3. The traditional grouting materials mainly use cement-based materials [67, 68]. However, the production process of cement involves high energy consumption, high CO_2 emissions, and the consumption of non-renewable mineral resources, which does not conform to the sustainable development concept of low-carbon and environmental protection. In addition, the hydration products of cement mainly consist of $Ca(OH)_2$ and C-S-H gel with a high Ca/Si ratio. The alkaline substances such as $Ca(OH)_2$ are prone to dissolve in water-rich conditions and will further lead to the decomposition of C-S-H gel, resulting in poor structural stability, water erosion resistance, and durability of cement-based grouting materials. On the other hand, existing studies have relatively conservative requirements for the performance indicators of the slurry, focusing more on the workability of the slurry during construction (setting time, consistency, and fluidity), and not paying enough attention to the performance of the slurry after hardening (strength, impermeability, and durability). To extend the setting time of the slurry and improve its fluidity, most grouting materials have low early strength, insufficient strength in the later stage, and poor durability.

1.3 Research Content and Technical Approach

1.3.1 Research Content

1.3.1.1 Research on the Activation Method of Electrolytic Manganese Slag

Taking the electrolytic manganese slag produced by a factory in Songtao County, Tongren City, Guizhou Province as the research object, the gelation activity of the electrolytic manganese slag was stimulated by high-temperature calcination. The effects of different temperatures on the mineral phases and chemical components of the electrolytic manganese slag were revealed through XRD and XRF tests. The thermal activation process of the electrolytic manganese slag was analyzed by thermogravimetric and differential thermal analysis, and the optimal calcination temperature of the electrolytic manganese slag was determined based on the strength indicators.

1.3.1.2 Preparation and Performance Study of Electrolytic Manganese Slag-Based Grouting Materials

Determine the influence of raw material ratios, water-cement ratios, admixtures, etc. on the main performance indicators such as setting time, fluidity and strength of the electrolytic manganese slag-based grouting material; clarify the influence laws of raw material ratios, water-cement ratios, admixtures, etc. on the rheological properties, strength properties and volume stability of the electrolytic manganese slag-based grouting material.

1.3.1.3 Research on the Hydration Mechanism of Electrolytic Manganese Slag-Based Grouting Materials Prepared Through Multi-Source Industrial Solid Waste Co-Processing

The physical and chemical properties of electrolytic manganese slag were analyzed to reveal its basic grouting activity and the mechanism of its activation, to obtain the synergistic mode between electrolytic manganese slag and other types of solid waste, and to determine the composition design method of the electrolytic manganese slag-based grouting material.

1.3.1.4 Research on the Durability and Environmental Safety of Electrolytic Manganese Slag-Based Grouting Materials

Through tests on the chloride ion erosion and sulfate erosion of the electrolytic manganese slag-based grouting material, as well as by using various modern testing methods such as XRD, SEM, and MIP, the strength variation patterns of the electrolytic manganese slag-based grouting material under different erosion environments were clearly identified, and the microscopic change mechanisms of the electrolytic manganese slag-based grouting material under various erosion conditions were revealed. Through toxicity leaching and alkali leaching tests on the manganese slag-based grouting material, the solidification mechanisms of heavy metals and alkali components in the manganese slag-based grouting material were clarified.

1.3.2 Research Route

This research mainly focuses on the alkaline-activated grouting materials and conducts three aspects of work: Firstly, using XRD, XRF, TG, and strength tests, by setting different calcination temperatures, the optimal calcination temperature for activating the EMR activity is determined; secondly, using EMR, GGBS (ground

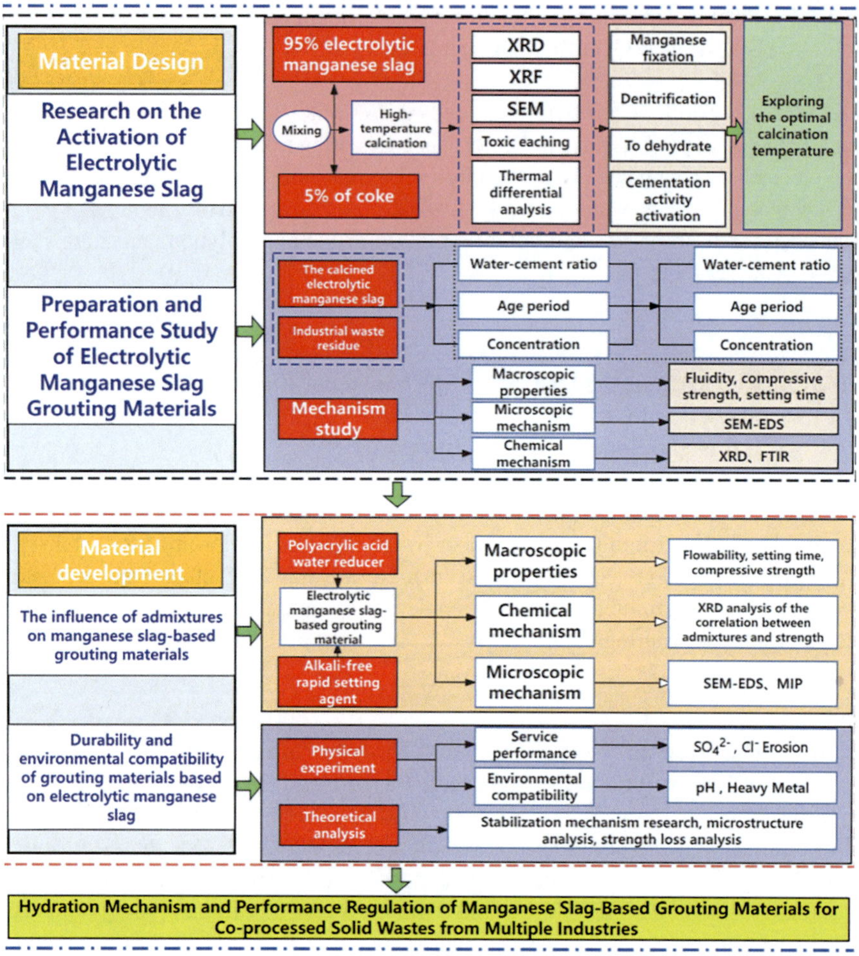

Fig. 1.4 Technology road map

granulated blast furnace slag) as raw materials and NaOH as the alkaline activator, three types of grouting materials (EFGM, ESGM, ECGM) are prepared. The influence of water-cement ratio and EMR content on their working performance and micro-structural characteristics (XRD, FTIR, SEM, MIP tests) is analyzed. At the same time, the effects of polycarboxylate superplasticizer and non-alkaline rapid-setting agent are studied; Thirdly, different concentration ionic solutions are prepared with $NaSO_4$ and NaCl. Through strength tests, the influence of ionic concentration and time on the material strength is investigated, and the intrinsic relationship between ionic erosion and strength is revealed.

According to the research plan, the specific technical roadmap of this study is shown in Figs. 1.4.

References

1. Feng, J. (2021). Problems and solutions in the treatment of industrial solid waste in China. *Leather Science and Engineering, 2*(18), 91–92.
2. Song, X., Liang, G., Luo, W., et al. (2021). Life cycle assessment of full industrial solid waste raw material cement process. *Acta Scientiae Circumstantiae, 41*(12), 5190–5199.
3. Gu, H., Zou, Z., Ye, T., et al. (2024). Current status and application trends of resource utilization of electrolytic manganese residue. *Modern Mining, 40*(4), 220–225.
4. Feng, Z., Zou, Z., Yang, Y., et al. (2023). Research on filling materials and properties of electrolytic manganese residue combined with steel slag. *Metal Mine, 12*, 245–251.
5. Shi, Y., Long, G., Wang, F., et al. (2023). Innovative co-treatment technology for effective disposal of electrolytic manganese residue. *Environmental Pollution, 335*, 122234.
6. Xie, X., Wang, F., Wang, G., et al. (2017). Progress in heavy metal pollution of surface water in China. *Environmental Science and Management, 42*(02), 31–34.
7. Chen, H., Long, Q., Zhou, F., et al. (2020). Elec-accumulating behaviors of manganese in the electrokinetics-processed electrolytic manganese residue with carbon dioxide and oxalic acid. *Journal Of Electroanalytical Chemistry, 865*, 114162.
8. Wang, F., Long, G., & Zhou, J. L. (2023). Deep insight into green remediation and hazard-free disposal of electrolytic manganese residue-based cementitious material. *Science of The Total Environment, 894*, 165049.
9. Liu, S., Xu, H., Wang, Q., et al. (2015). Current status of manganese ore supply and demand in China and suggestions for sustainable development. *Resources & Industries, 17*(06), 38–43.
10. Lin, T., Lu, Z., Wei, J., et al. (2013). Production technology and design of electrolytic manganese. *Tianjin Chemical Industry, 27*(03), 26–28.
11. Yan, W. (2008). Development of manganese ore resources and rich slag industry in China. *China Manganese Industry, 01*, 7–11.
12. Li, J., Li, J., Wei, H., et al. (2021). Alkaline-thermal activated electrolytic manganese residue-based geopolymers for efficient immobilization of heavy metals. *Construction And Building Materials, 298*, 123853.
13. Wang, Y., Cui, S., & Chi, B. (2013). Study on the influence of calcium oxide activity under different decomposition conditions. *Bulletin of The Chinese Ceramic Society, 32*(05), 829–832.
14. Huang, X., Zhao, X., Gao, Y., et al. (2022). Research on the setting and practice of solid waste treatment and disposal test projects based on professional background. *Guangdong Chemical Industry, 49*(23), 256–258.
15. Yi, G. (2024). Discussion on the development of shield tunnel construction technology and key management points. *Tunnel Construction, 44*(05), 927–942.
16. Tang, L., He, Z., Pei, S., et al. (2024). Hydration characteristics and pollutant solidification mechanism of full solid waste electrolytic manganese residue super sulfate cement. *Journal of Building Engineering, 84*, 108563.
17. Fu, Y., Qiao, H., Feng, Q., et al. (2023). Review of new methods for resource utilisation of electrolytic manganese residue and its application in building materials. *Construction and Building Materials, 401*, 132901.
18. Hou, P., Qian, J., & Wang, Z. (2012). Production of quasi-sulphoaluminate cementitious materials with electrolytic manganese residue. *Cement & Concrete Composites, 34*(2), 248–254.
19. Lei, J., Peng, B., Chai, L., et al. (2014). Preparation of high-ferrum calcium aluminate cement clinker by electrolytic manganese residue. *Journal of Materials and Metallurgy, 13*(04), 257–261.
20. Zhao, S., Han, F., & Wang, Y. (2017). Research on composite slag Portland cement clinker prepared by electrolytic manganese residue and steel slag. *Bulletin of the Chinese Ceramic Society, 36*(05), 1766–1772.
21. He, W., Li, R., Zhang, Y., et al. (2022). Synergistic use of electrolytic manganese residue and barium slag to prepare belite- sulphoaluminate cement study. *Construction and Building Materials, 326*, 126672.

22. Ye, S., Cheng, H., Yuan, X., et al. (2024). Application of electrolytic manganese residue in preparation of ceramic aggregate and concrete. *Inorganic Chemicals Industry, 56*(06), 127–132.
23. Xu, W., Yu, J., & Wang, H. (2024). The influence of manganese slag on the properties of ultra-high-performance concrete. *Materials, 17*(2), 197.
24. He, J., Sun, C., Hu, W., et al. (2024). Study on the effect of silica-manganese slag mixing on the deterioration resistance of concrete under the action of salt freezing. *Buildings, 14*(9), 2684.
25. Tang, B., Gao, S., Wang, Y., et al. (2019). Pore structure analysis of electrolytic manganese residue based permeable brick by using industrial CT. *Construction and Building Materials, 208*, 697–709.
26. Yun, Y., Xu, Z., Ning, L., et al. (2024). Research on properties and hydration mechanism of mine filling materials prepared by co-milling electrolytic manganese residue-phosphogypsum-fly ash-calcium hydroxide mixture. *Industrial Safety and Environmental Protection, 50*(12), 99–105.
27. Luo, W., Li, B., Yang, G., et al. (2024). Utilisation of electrolytic manganese residue as a sulphate activator in producing concrete blocks with high-volume fly ash. *Journal of Cleaner Production, 434*, 139813.
28. Zhou, Y., Huang, L., Yun, Y., et al. (2022). Experimental study on preparation of fired bricks from electrolytic manganese residue in Songtao area. *Multipurpose Utilization of Mineral Resources, 1*, 1–22.
29. Wang, J., Wang, H., et al. (2019). Research on compressive and flexural properties of unfired bricks from electrolytic manganese. *Non-metallic Mines, 42*(03), 13–15.
30. Wang, D., Wang, Q., & Xue, J. (2020). Reuse of hazardous electrolytic manganese residue: Detailed leaching characterization and novel application as a cementitious material. *Resources Conservation and Recycling, 154*, 104645.
31. You, X., Wang, J., Wang, H., et al. (2020). Effect of mixing water content on properties of unfired bricks from electrolytic manganese residue. *Bulletin of the Chinese Ceramic Society, 39*(10), 3311–3315.
32. Lan, J., Zhang, S., Mei, T., et al. (2021). Mechanochemical modification of electrolytic manganese residue: Ammonium nitrogen recycling, heavy metal solidification, and baking-free brick preparation. *Journal of Cleaner Production, 329*, 129727.
33. Cai, M. (2024). Application of electrolytic manganese residue in highway subgrade filling construction. *Transpoworld, 20*, 63–65.
34. Qian, J. S., Hou, P. K., Wang, Z., et al. (2012). Crystallization characteristic of glass-ceramic made from electrolytic manganese residue. *Journal of Wuhan University of Technology-Materials Science Edition, 27*(1), 45–49.
35. Wang, G., Li, Z., Ming, Y., et al. (2013). Preparation of recycled ceramic tiles by using electrolytic manganese residue-waste ceramic grinding fine powder. *Bulletin of the Chinese Ceramic Society, 32*(08), 1496–1501.
36. Han, Y., Cui, X., Lv, X., et al. (2018). Preparation and characterization of geopolymers based on a phosphoric-acid-activated electrolytic manganese dioxide residue. *Journal of Cleaner Production, 205*, 488–498.
37. Mi, B., Zhao, H., Lu, M., et al. (2024). Synthesis of electrolytic manganese slag-solid waste-based geopolymers: Compressive strength and mn immobilization. *Materials, 17*(6), 1431.
38. Zhan, X., Wang, L., Wang, L., et al. (2019). Enhanced geopolymeric co-disposal efficiency of heavy metals from MSWI fly ash and electrolytic manganese residue using complex alkaline and calcining pre-treatment. *Waste Management, 98*, 135–143.
39. Zengqi, Z. (2018). *Study on the hydration kinetics model of cement-slag composite cementitious materials*. Tsinghua University.
40. Junxiang, W. (2017). *Study on the regulation research on hydration reaction characteristics of slag-based cementitious materials activated by CaO*. Shandong University of Science and Technology.

41. Liu, J., Jin, L., & Sun, L. (Eds.). (1990). *Production and performance of adhesive materials: Cement, lime, gypsum*. Tongji University Press.
42. Zheng, W., Zou, M., & Wang, Y. (2019). Research progress on alkali-activated cementitious materials. *Journal of Building Structures, 40*(01), 28–39.
43. Huang, X., Wang, Y., Zhou, Y., et al. (2023). Study on the anti-seepage performance of polyacrylate composite electrolytic manganese slag cementitious materials. *Metal Mine, 01*, 283–288.
44. Zhang, X., Liu, F., Zhu, J., et al. (2021). Preparation and Characterization of Composite Cementitious Material Based on Electrolytic Manganese Slag-Phosphogypsum. *Bulletin of The Chinese Ceramic Society, 40*(05), 1610–1619.
45. He, D., Chen, M., Liu, H., et al. (2024). Preparation of activated electrolytic manganese residue-slag-cement ternary blended cementitious material: Hydration characteristics and carbon reduction potential. *Construction and Building Materials, 425*, 135990.
46. Ding, S., Shui, Z. H., Chen, W., et al. (2014). Properties of supersulphated phosphogysumslag cement (SSC) concrete. *Journal of Wuhan University of Technology-Materials Science Edition, 29*(1), 109–113.
47. Li, J., Sun, Z., Pang, M., et al. (2020). Study on Preparation and Durability of Phosphogypsum Composite Cementitious Foam Concrete. *Journal of Physics. Conference Series, 1646*(1), 12102.
48. Hua, S., Wang, K., Yao, X., et al. (2016). Effects of fibers on mechanical properties and freeze-thaw resistance of phosphogypsum-slag based cementitious materials. *Construction and Building Materials, 121*, 290–299.
49. Chen, P., Ma, B., Tan, H., et al. (2022). Improving the mechanical property and water resistance of β-hemihydrate phosphogypsum by incorporating ground blast-furnace slag and steel slag. *Construction and Building Materials, 344*(Aug.15), 128261–128265.
50. Wu, Z., Feng, Z., Pu, S., et al. (2024). Mechanical properties and environmental characteristics of the synergistic preparation of cementitious materials using electrolytic manganese residue , steel slag, and blast furnace slag. *Construction and Building Materials, 411*, 134480.
51. Zhou, Y., Xie, L., Kong, D., et al. (2022). Research on optimizing performance of desulfurization-gypsum-based composite cementitious materials based on response surface method. *Construction and Building Materials, 341*, 127874.
52. Xie, F., Liu, Z., Zhang, D., et al. (2020). The effect of NaOH content on rheological properties, microstructures and interfacial characteristic of alkali activated phosphorus slag fresh pastes. *Construction And Building Materials, 252*, 8.
53. Hajimohammadi, A., & van Deventer, J. S. J. (2016). Dissolution behaviour of source materials for synthesis of geopolymer binders: A kinetic approach. *International Journal of Mineral Processing, 153*, 80–86.
54. Qin, P. (2019). Research progress on grouting technology for unfavorable geology. *Metal Mine, 06*, 1–5.
55. Ren, J., Xu, Y., Huang, J., et al. (2021). *Gradation optimization and strength mechanism of aggregate structure considering macroscopic and mesoscopic aggregate mechanical behaviour in porous asphalt mixture* (Vol. 300, p. 124262). Construction and Building Materials.
56. Han, X., Chen, M., & Jiao, H. (2018). Research status and prospect of polymer-cement composite grouting materials. *Bulletin of the Chinese Ceramic Society, 37*(08), 2457–2462.
57. Dong, X., & Wang, Y. (2023). Preparation Method and Research of Industrial Solid Waste-Based Anti-Seepage Grouting Materials. *Coal and Chemical Industry, 46*(11), 94–96.
58. Xiao, J., Liu, C., & He, J. (2023). Research on characteristics of grouting materials for synergistic grouting of multi-source solid wastes with CFB Ash and steel slag fines. *China Journal of Highway and Transport, 36*(08), 162–171.
59. Jing, W. (2023). *Research on Preparation and Performance of Industrial Solid Waste Modified Polymer Grouting Materials*. Wuhan University of Light Industry.
60. Tamburini, S., Natali, M., Garbin, E., et al. (2017). Geopolymer matrix for fibre reinforced composites aimed at strengthening masonry structures. *Construction and Building Materials, 141*, 542–552.

61. Song, W., Zhu, Z., Pu, S., et al. (2020). Synthesis and characterization of eco-friendly alkali-activated industrial solid waste-based two-component backfilling grouts for shield tunnelling. *Journal of Cleaner Production, 266*, 6848.
62. Jiang, B., Wu, M., Wu, S., et al. (2023). A review on development of industrial solid waste in tunnel grouting materials: Feasibility, performance, and prospects. *Materials, 16*(21), 6848.
63. Wang, Y., Meng, X., & Cheng, D. (2019). Research on the revision of the Law on the Prevention and Control of Environmental Pollution by Solid Waste. *Chinese Journal of Environmental Management, 11*(06), 90–94.
64. He, C., Wang, S., Liu, J, et al. *Design method of shield tunnel segment considering performance deterioration of Dual-grout synchronous grouting*
65. Han, Q., Yi, X., Chen, Y., et al. (2015). Experimental Study on Influence of Admixtures on Performance of Double-fluid Grout. *Journal of Chongqing Jiaotong University (Natural Science), 34*(01), 40–43.
66. Liu, J., Zhang, G., Yang, D., et al. (2016). Influence of Clay-Cement Slurry Rheological Properties on Grouting. *Bulletin of The Chinese Ceramic Society, 35*(08), 2581–2589.
67. Pelizza, S. P. D. B. L. (2010). Analysis of the performance of two component back-filling grout in tunnel boring machines operating under face pressure. *Politecnico Di Torino Repository Istituzionale, 54*, 435–447.
68. Hyett, A. J., Bawden, W. F., & Reichert, R. D. (1992). The effect of rock mass confinement on the bond strength offully grouted cable boltsfully grouted cable bolts. *International Journal of Rock Mechanics and Mining Sciences, 29*(5), 503–524.

Part I
Study on Pretreatment and Modification of Electrolytic Manganese Slag

Chapter 2
Research on Desulfurization Denitrification and Activation of Gelling Activity of Electrolytic Manganese Slag

2.1 Introduction

The chemical composition of electrolytic manganese slag is similar to that of the mineral phase composition of cementitious materials, laying the foundation for the preparation of cementitious materials from electrolytic manganese slag. This chapter first analyzes the physical and chemical properties of electrolytic manganese slag after water washing and ammonia removal. The use of calcination to treat electrolytic manganese slag can achieve efficient ammonia and sulfur dioxide removal, as well as the dissolution of heavy metals, while also making the electrolytic manganese slag after high-temperature calcination have good cementitious activity. By synergistically utilizing it with fly ash, desulfurization gypsum, and Portland cement clinker, high-performance manganese slag-based grouting materials are prepared to solve the problem of low comprehensive utilization rate of electrolytic manganese slag.

2.2 Physical and Chemical Properties of Electrolytic Manganese Slag

Understanding the basic physical and chemical properties of electrolytic manganese slag is the foundation for its resource utilization. This chapter analyzes the basic physical and chemical characteristics of electrolytic manganese slag through means such as XRF fluorescence analysis, X-ray diffraction analysis (XRD), and thermal difference analysis. It mainly includes the physical properties, chemical composition, and mineral phase composition of electrolytic manganese slag. This is to design reasonable experimental schemes for the thermal activation treatment and chemical activity stimulation of electrolytic manganese slag, and lay the foundation

Z. Wu et al., *Multi-Source Industrial Solid Waste Synergistically Utilized in Manganese Slag-Based Grouting Materials*,
https://doi.org/10.1007/978-981-92-3021-1_2

for the research and development of manganese slag-based grouting materials. The XRF fluorescence analysis experiment reveals the influence laws of different thermal activation treatment temperatures on the chemical composition of electrolytic manganese slag; the X-ray diffraction analysis (XRD) reveals the influence laws of different thermal activation temperatures on the mineral phase composition of electrolytic manganese slag; through the thermal difference analysis experiment, the variation laws of the weight loss rate of electrolytic manganese slag under different calcination temperatures are revealed, and the influence of temperature on the desulfurization rate of electrolytic manganese slag is verified by comparing with the results of the XRF fluorescence analysis experiment and the X-ray diffraction analysis (XRD) experiment.

2.2.1 *Physical Properties*

The electrolytic manganese slag used was sourced from Songtao, Guizhou Province. The electrolytic manganese slag was washed with water and left to naturally pile up for more than half a year. Due to the presence of a large amount of semi-hydrated gypsum in the electrolytic manganese slag, it has a certain adhesive effect on surrounding substances. After weathering for some time, it hardens into blocks. At the same time, the electrolytic manganese slag is composed of fine particles. During the stacking process, the particles agglomerate and also harden into blocks. The electrolytic manganese slag has a high moisture content, reaching 30%, mainly because the electrolytic manganese slag has a porous structure and strong water absorption. It needs to be dried before other experimental studies can be conducted. Therefore, the electrolytic manganese slag is dried, ground, and passed through a 0.075 mm sieve. Unheated activated electrolytic manganese slag has extremely low hydration and gelation activities and is not suitable as raw material for grouting materials, but activated electrolytic manganese slag through high-temperature activation shows good gelation activity and can be used as raw material to develop manganese slag-based grouting materials.

2.2.2 *Chemical Composition*

The chemical composition of the washed electrolytic manganese slag is shown in Table 2.1. As can be seen from the table, the contents of SiO_2 and gypsum are the highest in the electrolytic manganese slag, followed by CaO and Al_2O_3. In addition, it also contains pyrite and other metal compounds, as well as a large amount of SO_3.

Table 2.1 Determination of chemical components by X-ray fluorescence spectrometry

Material Name	Chemical components					
	SiO_2	CaO	Al_2O_3	Fe_2O_3	MgO	SO_3
Raw materials	31.57	18.12	7.73	4.37	2.76	35.3

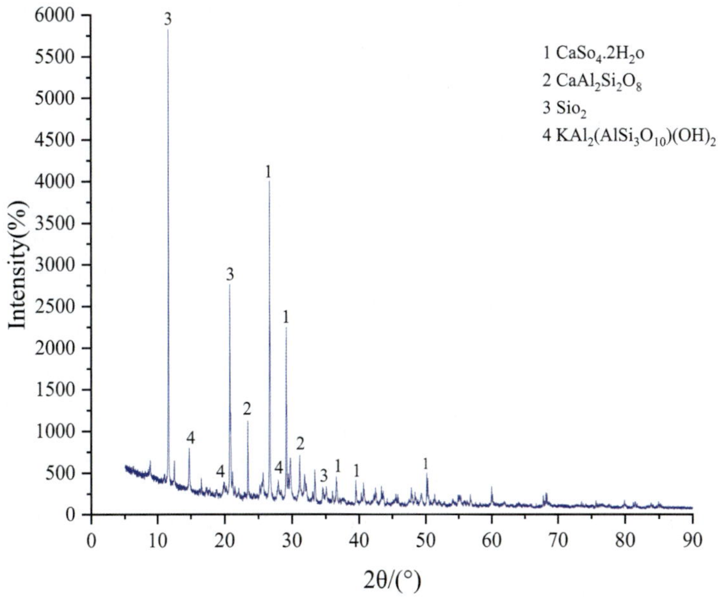

Fig. 2.1 XRD diffraction spectrum of electrolytic manganese slag

2.2.3 *Mineral Phase Composition*

The mineral phases of uncalcined electrolytic manganese slag are shown in Fig. 2.1. As can be seen from the figure, the main minerals of the electrolytic manganese slag are quartz, calcium sulfate dihydrate, calcium feldspar, muscovite, etc.

2.3 Research on Desulfurization, Denitrification, and Gelation Activity Activation of Electrolytic Manganese Slag

2.3.1 *Manganese Slag Thermal Activation Treatment*

This paper selects the washed desulfurized electrolytic manganese slag from Songtao County, Tongren City, Guizhou Province as the research object. The electrolytic manganese slag is treated with water washing to achieve desulfurization and removal of soluble salt substances. The washed electrolytic manganese slag mainly consists of insoluble $CaSO_4•2H_2O$ and SiO_2, and contains a large amount of SO_3, featuring high moisture content, high viscosity and low activity [1–3]. This paper realizes desulfurization and activation of activity of the electrolytic manganese slag through thermal activation treatment [4]. After high-temperature calcination up to 1000 °C, the main thermal activation reactions in the system are:

$$CaO + Al_2O_3 \rightarrow CaO{\cdot}Al_2O_3\left(CA\right) \tag{2.1}$$

$$2CaO + SiO_2 \rightarrow 2CaO{\cdot}SiO_2\left(C_2S\right) \tag{2.2}$$

The electrolytic manganese slag was crushed into blocks and placed in a 105 °C blast-drying oven for 4 h until a constant weight was achieved. The dried electrolytic manganese slag was then ground into powder using a grinding ball machine and could pass through a 0.075 mm square hole sieve. 5% coke was added to the ground electrolytic manganese slag as a reducing agent and was sent into a muffle furnace to be heated to the set temperature and properly maintained for desulfurization [5]. To study the influence laws of temperature on the desulfurization rate and activity excitation of electrolytic manganese slag, this paper set the desulfurization temperatures as 600 °C, 700 °C, 800 °C, 900 °C, 1000 °C, and 1100 °C, with a holding time of 30 min. After furnace cooling, different thermal activation temperature-treated calcined electrolytic manganese slag was obtained. The calcination process flow chart of electrolytic manganese slag is shown in Fig. 2.2. The effects of temperature increase on the desulfurization rate of electrolytic manganese slag were determined by thermogravimetric analysis, X-ray diffraction analysis and X-ray fluorescence spectrometer.

Figure 2.3 shows the thermogravimetric/differential thermal analysis diagram of electrolytic manganese slag from 0 °C to 800 °C. In the figure, TG represents thermogravimetric and DTA represents differential thermal analysis. It can be seen from the figure that as the temperature increases, the thermogravimetric curve gradually decreases, and the mass of the electrolytic manganese slag gradually decreases. This is mainly manifested as the sulfur contained in the electrolytic manganese slag being reduced to SO_2. When the temperature rises to 0 °C–100 °C, the decrease is

Fig. 2.2 Schematic diagram of high-temperature calcination of electrolytic manganese slag

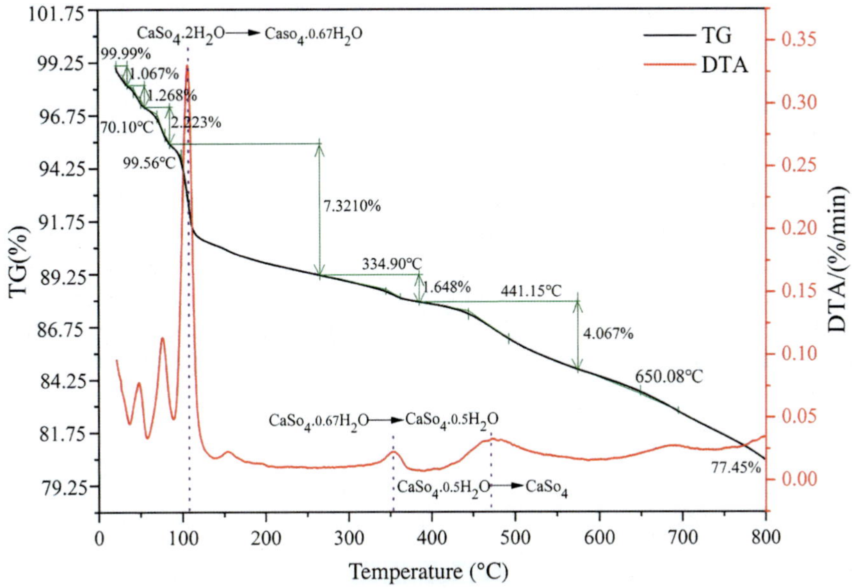

Fig. 2.3 Thermogravimetric/thermal difference analysis of electrolytic manganese slag 0–800 °C

Table 2.2 Determination of chemical components by X-ray fluorescence spectrometry

	Chemical components					
Material name	SiO_2	CaO	Al_2O_3	Fe_2O_3	MgO	SO_3
Raw materials	31.57	18.12	7.73	4.37	2.76	35.3
900 °C	37.85	23.98	12.86	10.435	8.19	3.67

the most obvious. When the temperature reaches 800 °C, the mass loss of the electrolytic manganese slag reaches 22.55%, indicating that the desulfurization rate is close to 77.45%. According to the DTA analysis, when the temperature rises to 110 °C, $CaSO_4{\bullet}2H_2O$ dehydrates to form β-$CaSO_4{\cdot}0.67H_2O$. The intuitive manifestation in the figure is the formation of a huge endothermic peak, that is, $CaSO_4{\bullet}2H_2O$ absorbs a large amount of heat for the conversion to β-$CaSO_4{\cdot}0.67H_2O$. When the temperature continues to rise to 350 °C, β-$CaSO_4{\cdot}0.67H_2O$ in the electrolytic manganese slag converts to $CaSO_4{\cdot}0.5H_2O$, that is, the gradual formation of crystalline calcium sulfate. However, the endothermic peak in the DTA curve is not particularly obvious. The main reason is that the transformation of calcium sulfate crystals is a gradual process. When the temperature reaches 470 °C, $CaSO_4{\cdot}0.5H_2O$ completely converts to soluble $CaSO_4$. As the temperature continues to rise to 680 °C, soluble $CaSO_4$ converts to insoluble $CaSO_4$ [6].

Table 2.2 shows the main chemical components of manganese slag raw material and electrolytic manganese slag after 900 °C calcination. After the electrolytic manganese slag undergoes 900 °C high-temperature activation treatment, the desulfurization rate is approximately 89.6%. Under the heat activation treatment, the structure of silicon and aluminum components in the manganese slag changes. The stable silicon-oxygen tetrahedral structure and aluminum-oxygen tetrahedral structure in the manganese slag transform into metastable glassy aluminum-silicon structure, which has a relatively high potential cementitious activity. When the temperature reaches 900 °C, the Ca, Al, and Si in the system undergo solidification reactions to form CA, C_2S, and C_3S.

Figure 2.4 shows the change patterns of mineral phases in electrolytic manganese slag at different high temperatures. From the figure, it can be seen that as the temperature increases, the $CaSO_4 \cdot 2H_2O$ in the electrolytic manganese slag undergoes four transformation processes. Firstly, $CaSO_4 \cdot 2H_2O$ undergoes dehydration to form β-$CaSO_4 \cdot 0.67H_2O$, and then β-$CaSO_4 \cdot 0.67H_2O$ transforms into $CaSO_4 \cdot 0.5H_2O$; subsequently, all $CaSO_4 \cdot 0.5H_2O$ is converted into soluble $CaSO_4$; as the temperature continues to rise, the soluble $CaSO_4$ transforms into insoluble $CaSO_4$. The diffraction peak of $CaSO_4 \cdot 0.5H_2O$ in the XRD analysis diagram of the electrolytic manganese slag corresponding to the figure gradually decreases with the increase in temperature. Through the analysis of the XRD diffraction analysis diagrams of electrolytic manganese slag at different temperatures, the experimental results of the thermal difference analysis test are proved. Comparing the characteristic peak positions of the manganese slag raw material and the spectrum at 600 °C, when the calcination temperature reaches 600 °C, the characteristic peak of $CaSO_4 \cdot 2H_2O$ completely disappears, and the characteristic peak of $CaSO_4$ appears. Comparing the characteristic peak positions of the spectra at 600 °C, 700 °C, 800 °C, 900 °C, and 1000 °C, it is basically consistent, proving that after the temperature reaches 470 °C, all $CaSO_4 \cdot 0.5H_2O$ is completely converted into anhydrous $CaSO_4$.

2.3.2 The Influence of Thermal Activation Treatment on the Activity of Manganese Slag

The main components of electrolytic manganese slag are SiO_2 and $CaSO_4$. In the raw material of uncalcined electrolytic manganese slag, $CaSO_4$ exists in the form of dihydrate, but it will convert to an anhydrous form under high-temperature calcination. This was confirmed through XRD and differential thermal analysis tests. $CaSO_4$ dissolves in water and ionizes into Ca^{2+} and SO_4^{2-} ions, which then react with the silicate cement. At the same time, SiO_2 and Al_2O_3, which have potential gelation activity, also undergo hydration reactions, providing mechanical strength for the stone body [7]. This paper uses the strength ratio method to determine the influence law of different temperatures on the basic gelation activity of the manganese slag after thermal activation treatment, laying the foundation for subsequent research. The formula of the strength ratio method is as follows (2.3):

$$H = P / P_0 \tag{2.3}$$

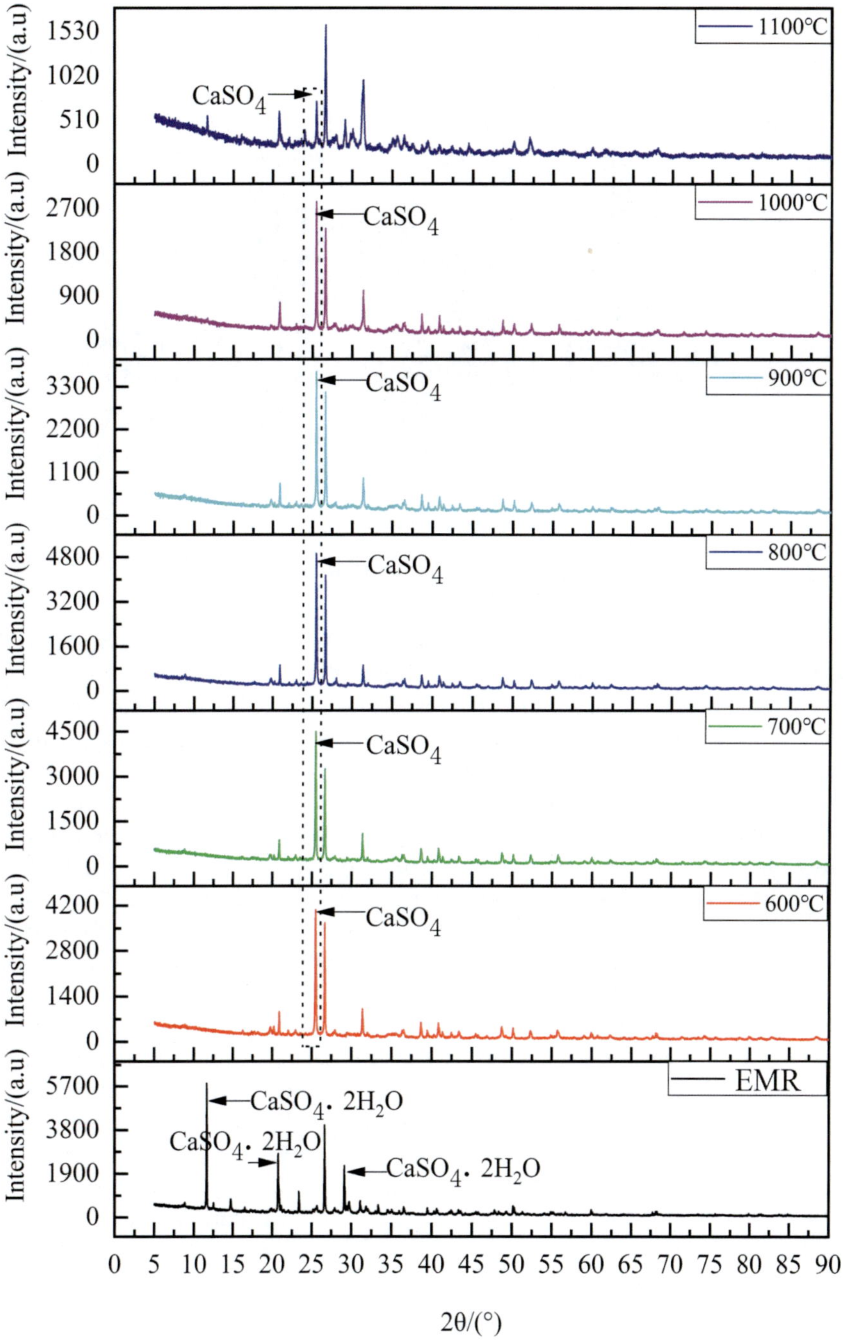

Fig. 2.4 XRD of electrolytic manganese slag at different temperatures

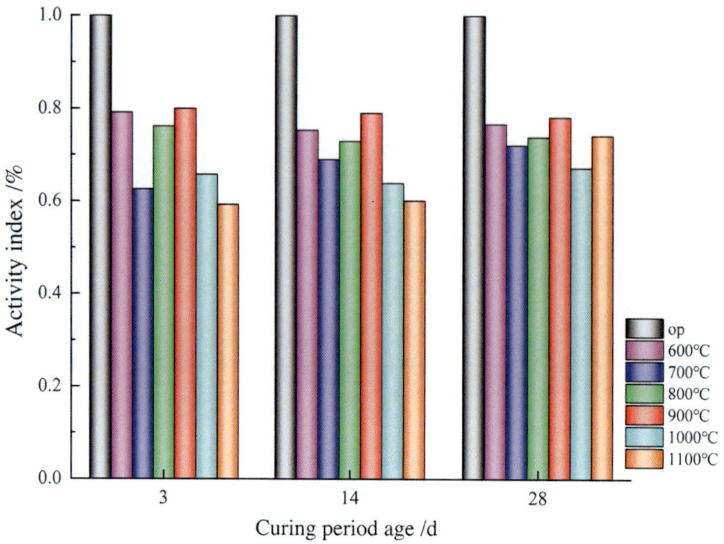

Fig. 2.5 Variation of activity index of manganese slag with temperature

In the formula: H represents the activity index of manganese slag after thermal activation treatment; *P* is the strength of the cement with 30% waste materials added during the curing period; P_0 is the strength of pure cement during the curing period.

First, use a precision balance with an accuracy of 0.05 g to accurately weigh the P42.5 silicate cement for the control group. Pour the weighed cement into the NJ-160 paste mixer and stir for 60 s until all the materials are evenly mixed. Then, measure the corresponding amount of tap water with a measuring cylinder and pour it into the mixer. Stir the mixer at a slow speed to prevent the paste from splashing during the mixing process. The stirring time is 120 s. Pour the pre-mixed paste into a 40 × 40 × 160 mm test mold and place the filled test mold on the vibration table. Vibrate it rapidly for 60 seconds to ensure that all bubbles are expelled. After 24 hours, remove the mold and place it in the curing box. Replace 30% of the silicate cement with equal amounts of 600 °C, 700 °C, 800 °C, 900 °C, 1000 °C, and 1100 °C calcined electrolytic manganese slag. Repeat the above steps to prepare the samples. The curing age of the samples is 3 days, 14 days, and 28 days, respectively.

Figure 2.5 shows the variation pattern of the activity index of manganese slag with temperature. From the figure, it can be seen that when the calcination temperature is 900 °C, the activity index of electrolytic manganese slag is the highest, reaching 0.8. 900 °C has the most significant effect on activating the basic gelation activity of electrolytic manganese slag. The activity index at 600 °C is second only to 900 °C. However, as can be seen from the TG curve in Figure 1, when the calcination temperature is 600 °C, the mass loss of manganese slag is approximately 17.6%. At this time, the desulfurization rate of manganese slag is low and does not meet the requirements of the admixture for grouting materials. The main reason for the differences in the activity index of electrolytic manganese slag at different temperatures is as follows: When manganese slag is calcined at 600 °C at high temperature, $CaSO_4$ is in a transitional state from hydrated calcium sulfate to anhydrous calcium sulfate. At this time, $CaSO_4$ is easily soluble in water and dissolves and ionizes a large number of Ca^{2+} and SO_4^{2-} ions. Therefore, the source of the gelation activity of electrolytic manganese slag under 600 °C high-temperature calcination is mainly provided by $CaSO_4$. As the calcination temperature increases, $CaSO_4$ gradually changes from being easily soluble in water to being less soluble in water, and the proportion of $CaSO_4$'s gelation activity gradually decreases. SiO_2 and Al_2O_3 gradually transform into silicate glass bodies with the increase in temperature, and the proportion of gelation activity gradually increases. This is the main source of the gelation activity of manganese slag under 900 °C high-temperature calcination.

2.3.3 The Influence of Alkaline Environment on the Cementing Activity of Manganese Slag

In this study, NaOH was selected as the alkaline activator, which was a transparent, semi-solid solid with a purity of no less than 98%. The NaOH content was 5%, 10%, 15%, and 20% of the electrolytic manganese slag content, and the water-cement ratio was 0.5. According to the sample preparation steps mentioned above, triplex 40 × 40 × 160 mm samples were prepared, and the curing ages were 3 days, 14 days, and 28 days.

From Figs. 2.6 and 2.7, it can be seen that as the amount of alkaline activator increases, the compressive and flexural strengths of the specimens at different curing ages show a trend of first increasing and then decreasing, indicating that NaOH has a significant stimulating effect on the cementing activity of electrolytic manganese slag. When the NaOH content is 15%, the compressive strengths of the stone bodies at 3d, 14d, and 28d all reach the maximum values. The main reason is that when the amount of alkaline activator is low, the concentration of OH^- ions in the slurry is low, and most of the potential SiO_2 and Al_2O_3 substances in the manganese slag cannot be stimulated. The hydration reaction process is slow, and the early strength is low. As the amount of alkaline activator increases, the concentration of ionized OH^- ions in the slurry rapidly increases. The active substances SiO_2 and Al_2O_3 in the manganese slag particles fully participate in the hydration reaction [8], and the amount of hydration products increases, resulting in a relatively faster increase in strength. A NaOH content of 15% is the best for stimulating the cementing activity of electrolytic manganese slag; when the amount of alkaline activator is high, the OH^- ions used to activate the silicon-aluminum components of the manganese slag rapidly ionize, and the hydration reaction occurs rapidly. The surfaces of unreacted manganese slag particles are quickly covered with a large amount of hydration and cementing products, hindering the internal hydration reaction of the particles, resulting in a relatively slow increase in strength [9]. In summary, the amount of alkaline activator has a significant effect on the stimulation of the

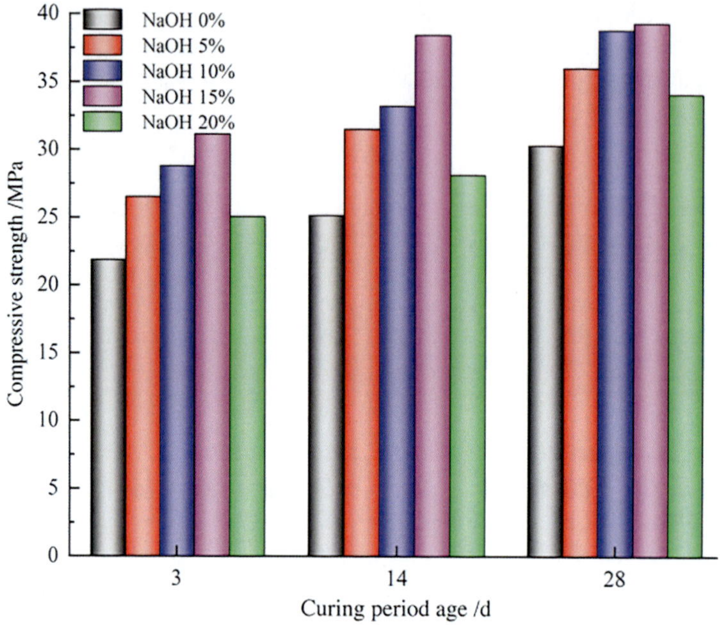

Fig. 2.6 Effect of NaOH on compressive Strength

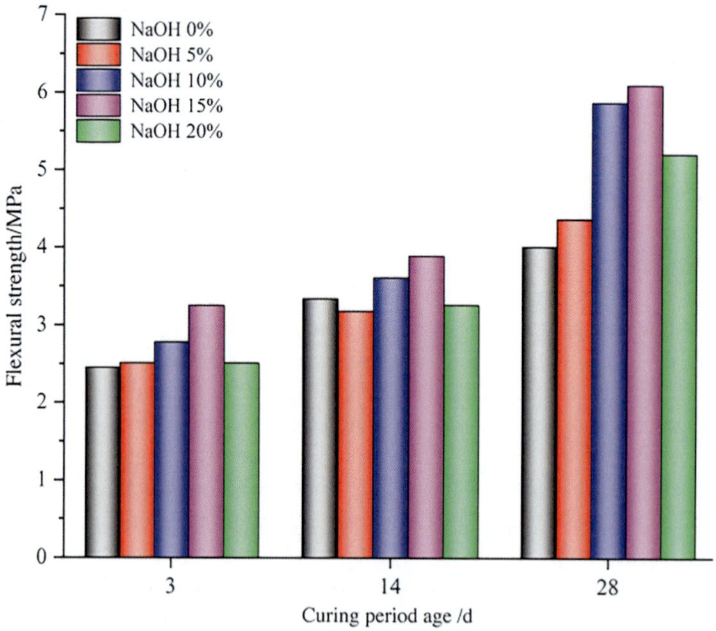

Fig. 2.7 Effect of NaOH on flexural Strength

cementing activity of manganese slag and has a significant impact on the mechanical properties of the stone.

2.4 Chapter Summary

This chapter reveals the influence laws of the thermal activation temperature on the desulfurization rate, chemical composition, and mineral phase composition of electrolytic manganese slag through thermogravimetric analysis, XRF diffraction analysis, and X-ray fluorescence spectrometer. The impact of thermal activation treatment and alkaline excitation on the cementing activity of electrolytic manganese slag was tested using the ratio method, and the results are summarized as follows:

1. Thermal differential-thermogravimetric analysis indicates that when the temperature reaches 800 °C, the mass loss rate of the electrolytic manganese slag reaches 22.55%, meaning the desulfurization rate is approximately 77.45%. Combined with XRF fluorescence spectrometer analysis, after the electrolytic manganese slag undergoes 900 °C high-temperature activation treatment, the desulfurization rate is approximately 89.6%.
2. Based on the XRD diffraction spectra, when the calcination temperature reached 600 °C,$CaSO_4 \cdot 2H_2O$ was converted to $CaSO_4$, and after the temperature reached 470 °C, all $CaSO_4 \cdot 0.5H_2O$ was completely transformed into anhydrous $CaSO_4$.
3. When the calcination temperature is 900 °C, the activity index of the electrolytic manganese slag is the highest, reaching 0.8. Therefore, 900 °C is regarded as the optimal temperature for activating and enhancing the gelation activity of the electrolytic manganese slag.
4. When the NaOH content is 15%, the potential active SiO_2 and Al_2O_3 substances are maximally activated, and the compressive strength reaches its peak at all curing ages.

References

1. Ding, S., Shui, Z. H., Chen, W., et al. (2014). Properties of supersulphated phosphogysumslag cement (SSC) concrete. *Journal of Wuhan University of Technology-Materials Science Edition, 29*(1), 109–113.
2. Li, J., Sun, Z., Pang, M., et al. (2020). Study on preparation and durability of Phosphogypsum composite cementitious foam concrete. *Journal of Physics. Conference Series, 1646*(1), 12102.
3. Hua, S., Wang, K., Yao, X., et al. (2016). Effects of fibers on mechanical properties and freeze-thaw resistance of phosphogypsum-slag based cementitious materials. *Construction and Building Materials, 121*, 290–299.
4. Chen, P., Ma, B., Tan, H., et al. (2022). Improving the mechanical property and water resistance of β-hemihydrate phosphogypsum by incorporating ground blast-furnace slag and steel slag. *Construction and Building Materials, 344*(Aug.15), 128261–128265.

5. Wu, Z., Feng, Z., Pu, S., et al. (2024). Mechanical properties and environmental characteristics of the synergistic preparation of cementitious materials using electrolytic manganese residue, steel slag, and blast furnace slag. *Construction and Building Materials, 411*, 134480.
6. Zhou, Y., Xie, L., Kong, D., et al. (2022). Research on optimizing performance of desulfurization-gypsum-based composite cementitious materials based on response surface method. *Construction and Building Materials, 341*, 127874.
7. Xie, F., Liu, Z., Zhang, D., et al. (2020). The effect of NaOH content on rheological properties, microstructures and interfacial characteristic of alkali activated phosphorus slag fresh pastes. *Construction and Building Materials, 252*, 8.
8. Hajimohammadi, A., & van Deventer, J. S. J. (2016). Dissolution behaviour of source materials for synthesis of geopolymer binders: A kinetic approach. *International Journal of Mineral Processing, 153*, 80–86.
9. Qin, P. (2019). Research progress on grouting technology for unfavorable geology. *Metal Mine, 06*, 1–5.

Part II
Basic Performance Evaluation of Manganese Slag-Based Grouting

Chapter 3
Performance Evaluation of Electrolytic Manganese Residue-Based Grouting Materials

3.1 Introduction

Electrolytic Manganese Residue (EMR) is an industrial by-product generated from manganese ore through acid leaching, neutralization, purification, and filter pressing in the wet-process production of electrolytic manganese metal. This chapter investigates the mechanism of thermal activation of EMR via high-temperature calcination using X-ray diffraction (XRD), X-ray fluorescence (XRF), and thermogravimetric (TG) analyses. The optimal calcination temperature for activating EMR is determined through compressive strength tests.

Using EMR, ground granulated blast-furnace slag (GGBS), fly ash (FA), steel slag (SS), and coal slag (CS) as the main raw materials, and sodium hydroxide (NaOH) as the alkaline activator, three distinct grouting material systems were prepared:

Electrolytic manganese residue–GGBS–fly ash (EFGM),
Electrolytic manganese residue–GGBS–steel slag (ESGM),
Electrolytic manganese residue–GGBS–coal slag (ECGM).

The effects of water-to-binder ratio and EMR content on the fresh properties of the EMR-based grouts were systematically studied. Compressive strength tests were conducted to evaluate the influences of water-to-binder ratio, EMR content, and curing age on the mechanical properties. Furthermore, microstructural characteristics were examined through XRD, Fourier transform infrared spectroscopy (FTIR), scanning electron microscopy (SEM), and mercury intrusion porosimetry (MIP) to elucidate the impacts of mix proportion and curing duration on the hydration products, pore structure, and morphology of the grouts.

Z. Wu et al., *Multi-Source Industrial Solid Waste Synergistically Utilized in Manganese Slag-Based Grouting Materials*,
https://doi.org/10.1007/978-981-92-3021-1_3

3.2 Preparation of EFGM: An EMR-Based Grouting Material Incorporating GGBS and Fly Ash

3.2.1 *Fresh Properties of the Grout*

3.2.1.1 Flowability

The effects of EMR content and water-to-binder ratio on the flowability of EFGM are shown in Fig. 3.1. As can be seen from the figure, at a constant water-to-binder ratio of 0.4, the flowability of EFGM decreased from 164 mm to 90 mm—a reduction of 74 mm (45.1%)—as the EMR content increased from 10% to 60%. Similarly, at a water-to-binder ratio of 0.55, the flowability dropped from 226 mm to 132 mm, corresponding to a decrease of 94 mm (41.6%) over the same range of EMR content. This indicates that increasing the EMR content leads to a declining trend in flowability, and the absolute reduction in flowability becomes greater at higher water-to-binder ratios [1]. On the other hand, with the EMR content fixed at 10%, an increase in the water-to-binder ratio from 0.4 to 0.55 raised the flowability from 164 mm to 226 mm, an increment of 62 mm. When the EMR content was 60%, the same increase in water-to-binder ratio improved the flowability from 90 mm to

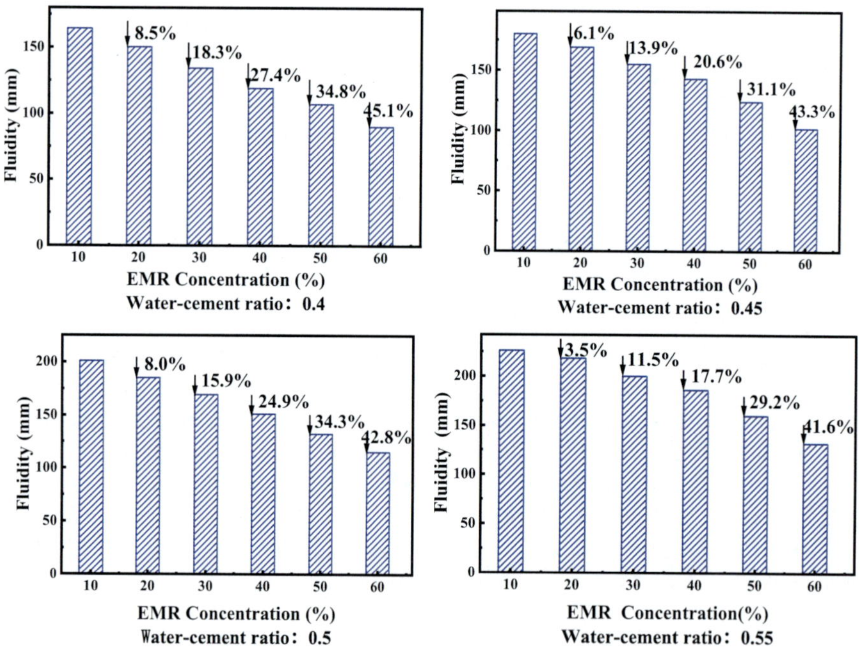

Fig. 3.1 Changes in fluidity with different manganese slag contents and different water-cement ratios

132 mm, an increase of 42 mm. In other words, a higher water-to-binder ratio generally enhances flowability, but the extent of this improvement diminishes as the EMR content increases.

EMR exhibits strong water absorption capacity [2]. At a constant water-to-binder ratio, an increase in EMR content leads to greater adsorption of free water in the system, thereby reducing the amount of available free water. This results in a decline in grout fluidity and, consequently, a reduction in flowability. Under a fixed EMR content, raising the water-to-binder ratio causes the water absorption of EMR to reach saturation, preventing further adsorption of free water. As a result, the content of free water in the system increases, which enhances the flowability of the grout [3, 4].

3.2.1.2 Initial and Final Setting Times

The initial and final setting times under different water-to-binder ratios and EMR contents are shown in Fig. 3.2. It can be observed that as the water-to-binder ratio increases, both the initial and final setting times exhibit a prolonging trend, and the extent of this prolongation becomes more pronounced with higher water-to-binder ratios. This is attributed to the increase in free water within the system at elevated water-to-binder ratios. Over the curing period, a significant portion of this free water

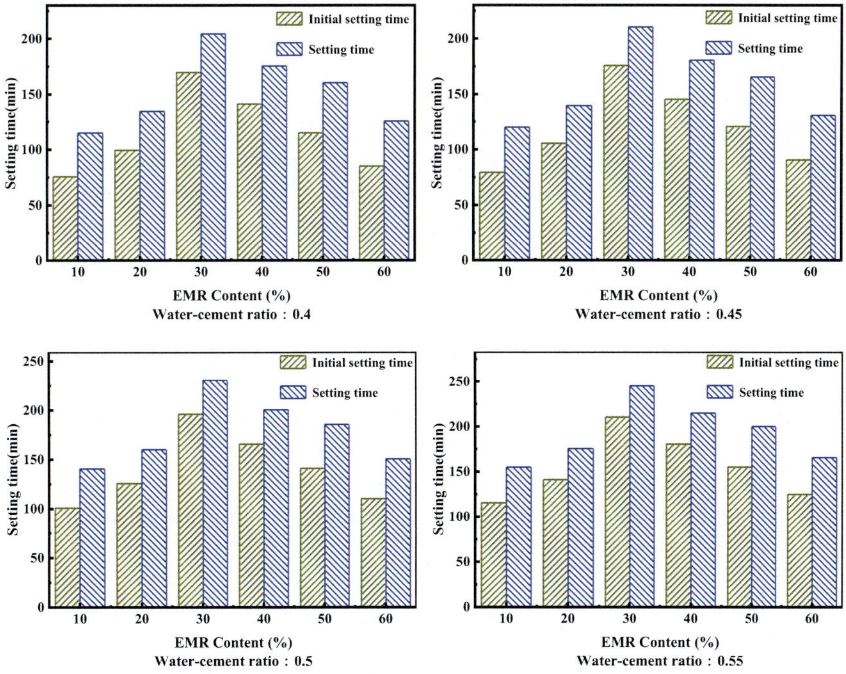

Fig. 3.2 Initial and final setting times

evaporates from the grout, thereby extending the setting time [5, 6]. With an increase in EMR content, the setting time of the grout first increases and then decreases, reaching its maximum when the EMR content is 30%. On one hand, the formation of geopolymers consumes free water in the system. A higher EMR content reduces the amount of available free water due to the water absorption of EMR, which in turn moderates the intensity of the geopolymerization reaction. On the other hand, an increase in EMR content corresponds to a decrease in GGBS content, leading to an overall reduction in reactive components and thus impeding the progress of geopolymerization. Furthermore, a higher EMR content introduces more calcium sulfate into the system, promoting the formation of substantial ettringite through hydration. The resulting ettringite can coat the surface of hydration products, hindering further hydration and consequently extending the setting time [7].

When the EMR content exceeds 30%, the increasing amount of calcium sulfate in the system leads to continued formation of ettringite. The persistent generation of ettringite introduces expansive stress, which accumulates and eventually disrupts the ettringite layer coating the reactants. This disruption allows the internal reactants to re-establish contact with surrounding water, resuming the hydration process and producing additional hydration products, consequently shortening the setting time [8].

3.2.2 Strength Characteristics of the Grouting Material

Figure 3.3 shows the compressive strength of EFGM with different water-to-binder ratios and EMR contents. As can be seen from the figure, the compressive strength of EFGM is inversely proportional to the water-to-binder ratio, and the extent of strength reduction increases as the water-to-binder ratio rises. When the water-to-binder ratio continuously increases while the concentration of the alkali activator remains constant, the dosage of the alkali activator increases accordingly. However, the amount of water molecules participating in the hydration reaction is relatively low. The excess water molecules hinder the hydration reaction of EFGM, leading to a continuous decrease in compressive strength. Moreover, the excessive water molecules in EFGM significantly increase the porosity of the specimens, further reducing the compressive strength.

When the EMR content exceeds 50%, the compressive strength of the material shows a significant declining trend. In contrast, when the EMR content is between 30% and 50% and the water-to-binder ratio ranges from 0.4 to 0.5, the variation in the 7-day and 28-day compressive strength is relatively moderate. During the mid-to-late stages of hydration (i.e., at 7 and 28 days), with a fixed water-to-binder ratio of 0.4, the compressive strengths reach 26.1 MPa and 34.9 MPa, respectively, at an EMR content of 20%. When the EMR content increases to 50%, the compressive strengths are 25.4 MPa and 32 MPa, respectively. At an EMR content of 60%, the compressive strengths drop to 15.4 MPa and 19.6 MPa, respectively. Since calcined EMR possesses high activity under the action of the alkali activator, and GGBS itself exhibits high activity [9], the highest compressive strength is achieved at an EMR content of 10%. As the EMR content increases (i.e., the GGBS content gradually decreases), the calcined EMR

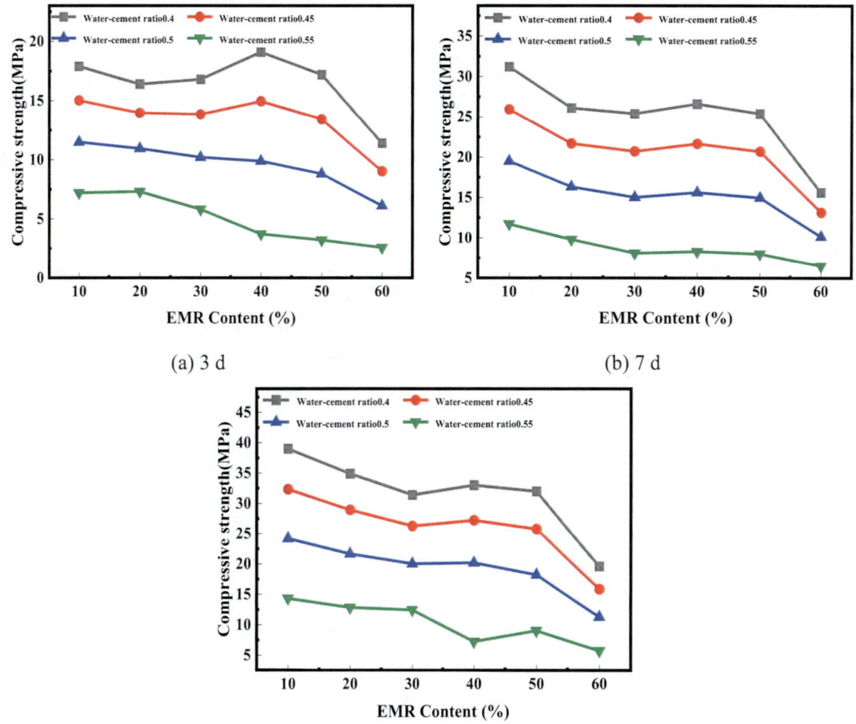

Fig. 3.3 Compressive strength of concrete with different manganese slag contents and different water-cement ratios at different ages. (**a**) 3 d. (**b**) 7 d. (**c**) 28d

maintains relatively high activity. Moreover, EMR exerts a certain sulfate activation effect on GGBS, which explains why the compressive strength of the specimens does not change significantly when the EMR content is within the range of 20% to 50%. However, when the EMR content increases to 60%, the GGBS content is only 10%. An excessively low GGBS content adversely affects the overall cementitious activity of EFGM. Although the excessive EMR can provide a certain level of strength through its own activity and some early-stage reactions, the strength remains relatively low in later curing stages and fails to reach the desired level. In summary, from the perspective of the key performance indicator of strength, the EMR content should not exceed 50%.

3.2.3 Microstructural Characteristics of the Grouting Material

3.2.3.1 XRD

Figure 3.4 shows the XRD patterns of samples with a water-to-binder ratio of 0.4 and a curing age of 28 days at different EMR contents. The detected hydration products primarily consist of C-(A)-S-H and N-A-S-H gels, along with unreacted quartz

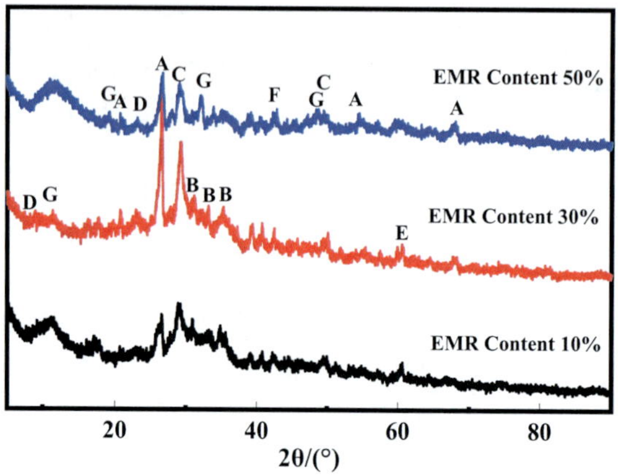

Fig. 3.4 XRD spectra analysis of samples with different EMR contents (A: Quartz—79-1906, B: Magnesite—79-1453, C: Calcium carbonate—72-1937, Calcium aluminosilicate—41-1451, E: Calcium hydroxide—87-0673, F: Calcium silicate—23-1042, G: Sodium sulfate—99-0103)

(SiO_2, PDF#79–1906), mullite ($3Al_2O_3{\cdot}2SiO_2$, PDF#79–1453), and other crystalline phases such as calcium carbonate ($CaCO_3$, PDF#72–1937), ettringite ($3CaO{\cdot}Al_2O_3{\cdot}CaSO_4{\cdot}32H_2O$, PDF#41–1451), portlandite ($Ca(OH)_2$, PDF#87–0673), calcium silicate (Ca_2SiO_4, PDF#23–1042), and sodium sulfate (Na_2SO_4, PDF#99–0103). Broad diffraction humps within the ranges of 5–10° and 20–30° are characteristic of the amorphous C-(A)-S-H and N-A-S-H gels, indicating the formation of a substantial amount of geopolymer gel during hydration.

At an EMR content of 30%, the diffraction peaks for quartz and calcium carbonate are the most intense, suggesting a higher amount of unreacted quartz and formed calcium carbonate. A small amount of residual ettringite is also observed. Furthermore, the diffraction peak of mullite appears in the 30–40° range, confirming that mullite from the fly ash was not fully consumed. Compared to patterns at other EMR contents, the consumption of mullite appears lower at this specific dosage. This phenomenon may be attributed to the varying proportions of raw materials resulting from different EMR contents, leading to differences in the quantities of unreacted phases remaining after hydration.

Figure 3.5 presents the XRD patterns of specimens with a water-to-binder ratio of 0.4 and an EMR content of 30% at different curing ages. As can be seen from the figure, the hydration products formed at different ages are largely similar. The main hydration products consist of C-(A)-S-H and N-A-S-H gels, along with unreacted phases such as quartz, mullite, calcium carbonate, ettringite, portlandite, calcium silicate, and sodium sulfate. The characteristic diffraction humps in the ranges of 5–10° and 20–30° correspond to the amorphous C-(A)-S-H and N-A-S-H gels. With increasing curing age, these diffraction features gradually intensify, indicating continuous formation of geopolymer gels. As the curing age progresses, the dissolution-polymerization process of the geopolymeric reaction proceeds, leading to a gradual enhancement of the quartz diffraction peak. Concurrently, the amount of calcium carbonate, formed from the reaction of atmospheric CO_2 with OH^- and Ca^{2+} in the system, also increases over time.

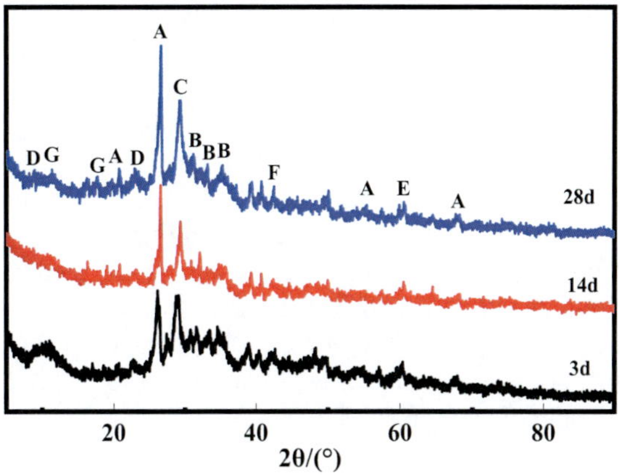

Fig. 3.5 XRD spectrum analysis of specimens with different curing ages (A: Quartz—79-1906, B: Magnesite—79-1453, C: Calcium Carbonate—72-1937, D: Calcium Aluminate—41-1451, E: Calcium Hydroxide—87-0673, F: Calcium Silicate—23-1042, G: Sodium Sulfate—99-0103)

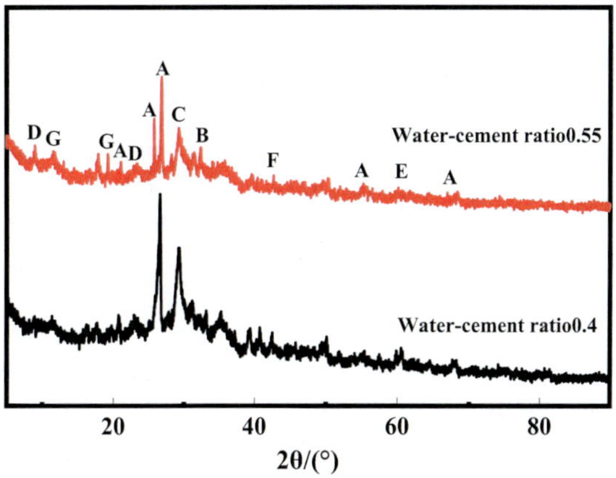

Fig. 3.6 XRD spectra analysis of samples with different water-cement ratios (A: Quartz—79-1906, B: Magnesite—79-1453, C: Calcium carbonate—72-1937, D: Calcium aluminosilicate—41-1451, E: Calcium hydroxide—87-0673, F: Calcium silicate—23-1042, G: Sodium sulfate—99-0103)

No distinct diffraction peaks of ettringite are observed at 3 and 14 days, whereas a small amount of ettringite is detected at 28 days. This may be attributed to the continuous consumption of ettringite formed during the early and middle stages of hydration, resulting in no prominent ettringite peaks in the initial and middle periods. By the later stage of hydration (28 days), as hydration products continue to form and the reactants that interact with ettringite are gradually depleted, a small amount of ettringite becomes detectable. Similarly, in the later hydration stage, unreacted SO_4^{2-} combines with Na^+ in the system to form sodium sulfate crystals, explaining the emergence of sodium sulfate diffraction peaks at later ages.

Figure 3.6 displays the XRD patterns of specimens with an EMR content of 30% and a curing age of 28 days under different water-to-binder ratios. The results indicate that the hydration products are essentially consistent across the varying water-to-binder ratios. The primary hydration products comprise C-(A)-S-H and N-A-S-H gels, alongside unreacted phases such as quartz and mullite, as well as crystalline hydration products including calcium carbonate, ettringite, portlandite, calcium silicate, and sodium sulfate. The characteristic diffraction humps in the ranges of 5–10° and 20–30° correspond to the C-(A)-S-H and N-A-S-H gels. As the water-to-binder ratio increases, the intensity of these diffraction features weakens, suggesting a reduction in the formation of cementitious gels. This observation aligns with the trend observed in the strength development. At a water-to-binder ratio of 0.55, distinct diffraction peaks of ettringite appear in the hydration products, indicating less consumption of ettringite under higher water-to-binder conditions. Concurrently, a quartz diffraction peak is observed at a diffraction angle of 25.74°, signifying that this crystalline quartz phase remains largely unreacted when the water-to-binder ratio is elevated. Furthermore, the intensity of the sodium sulfate diffraction peak increases at higher water-to-binder ratios, pointing to the generation of a greater amount of sodium sulfate.

3.2.3.2 FTIR

Figure 3.7 presents the FTIR spectra of the NaOH-activated electrolytic manganese residue–slag–fly ash grouting material. As shown in Figure 3.7a, the absorption bands near 3449 cm^{-1} and 3456 cm^{-1} broaden with increasing EMR content, indicating that the dissolution of Ca^{2+} from the EMR promotes the transformation of hydroxyl association states, which is consistent with the variation in $Ca(OH)_2$ content detected by XRD. The characteristic peak at 1645 cm^{-1} corresponds to the H–O–H bending vibration, and its intensity decreases with higher EMR content. This suggests a competitive interaction between the sulfate components in the EMR [10] and the pore water in the gel, leading to coarsening of the pore structure. The asymmetric stretching vibration at approximately 1433 cm^{-1} (attributed to CO_3^{2-}) may originate from the carbonation of sodium carbonate in calcite or the activator.

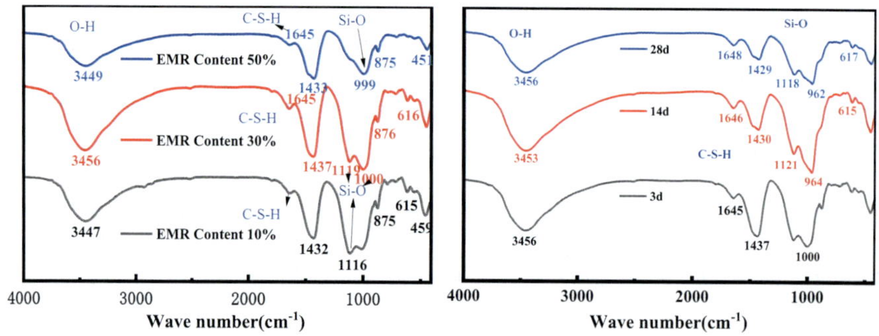

(a) At different levels of EMR content

(b) Samples with 30% EMR content and different curing days

Fig. 3.7 Infrared Spectrum of EMR—Slag—Fly Ash Grouting Material (**a**) At different levels of EMR content. (**b**) Samples with 30% EMR content and different curing days

It is noteworthy that the shift of the absorption band around 1000 cm^{-1} in the hardened samples is likely related to the stretching vibrations of Si–O–Si/Si–O–Al, which belong to the silicate structures in slag or fly ash, indicating the formation of geopolymer gels during the geopolymerization process. As the EMR content increases, the rising Ca/Si ratio accelerates the depolymerization rate of silicate tetrahedra, promoting the formation of more C–S–H gel. The band near 999 cm^{-1} may correspond to the Si–O vibration in C–S–H gel. With increasing EMR content, this band also broadens, suggesting that SiO_2 mainly originates from the EMR. Meanwhile, the bending vibration of the $[AlO_4]^{5-}$ tetrahedron appears near 615 cm^{-1}, and its absorption wavelength decreases with higher EMR content, revealing a change in the coordination state of the aluminate structure and the formation of a more stable Al-modified gel phase.

Figure 3.7b shows the FTIR spectra of the prepared grouting material at different curing ages. With prolonged curing, the intensity of the O–H vibration band near 3456 cm^{-1} decreases, which is attributed to the reaction between dissolved Ca^{2+} from the EMR and active SiO_2 from the slag, generating highly polymerized C–S–H gel phases and increasing the content of structurally bound water. In contrast, the characteristic band near 1648 cm^{-1} shows a reverse strengthening trend, which is related to the higher porosity of the gel framework formed at early ages. The ongoing hydration during later curing stages leads to a concentration of pore size distribution in the mesoporous range, enhancing capillary adsorption. The decreased intensity of the carbonate band near 1429 cm^{-1} is due to the dense coating formed by the newly generated C–S–H gel in the later curing stage, effectively inhibiting carbonation of the grouting material. The dynamic change of the characteristic band near 961 cm^{-1}, which appears at 14 days, corresponds to the transformation of C–S–H gel from a low to a medium degree of polymerization. The weakened intensity at 28 days reflects further reorganization of the hydrated cementitious network, forming an ordered C–S–H-type structure with a high Ca/Si ratio, accompanied by a shift of the Si–O vibration mode to the range of 930–940 cm^{-1}.

3.2.3.3 SEM

Figure 3.8 shows the micromorphology of specimens with a water-to-binder ratio of 0.4 and a curing age of 28 days at different EMR contents. The newly developed grouting material exhibits the formation of abundant flocculent gels. As the EMR content increases, the porosity of the specimens begins to rise, and the structure gradually becomes looser.

At an EMR content of 10%, under a magnification of 20,000×, a dense flocculent gel intermixed with needle-like ettringite crystals can be observed, contributing to a more compact microstructure. When the EMR content reaches 30%, both flocculent gels and pores are clearly visible. At an EMR content of 50%, the structure becomes further loosened with enlarged pores, which aligns with the conclusions drawn from XRD and MIP analyses.

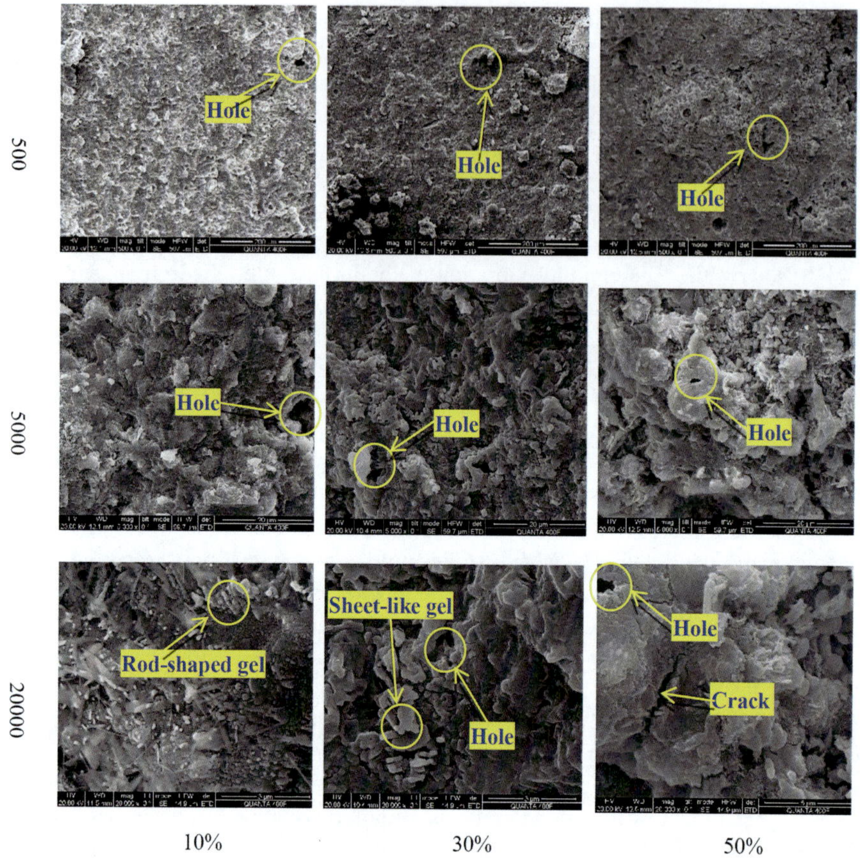

Fig. 3.8 Microscopic morphology of samples with different EMR contents

Figure 3.9 presents the micromorphology of specimens with a water-to-binder ratio of 0.4 and an EMR content of 30% at different curing ages. It can be observed that as the curing age increases, the amount of flocculent gel formed becomes greater, and the structure becomes increasingly denser, which is consistent with the trend in strength development.

At a curing age of 3 days, the specimen exhibits a relatively compact structure with few pores and a substantial amount of hydration products, corresponding to a grout strength of 16 MPa. Under a magnification of 5000× at 14 days, interconnected microcracks can be seen in the specimen. These microcracks may have been induced during sample preparation when the specimen was fractured [11], or they could be the result of expansive stress generated during the polymerization of unreacted materials encapsulated by rapidly formed gel [8]. When the curing age reaches 28 days, under a magnification of 20,000×, it is observed that the flocculent gels interweave to form a dense and integrated microstructure.

Figure 3.10 shows the micromorphology of specimens with an EMR content of 30% and a curing age of 28 days under different water-to-binder ratios. As the

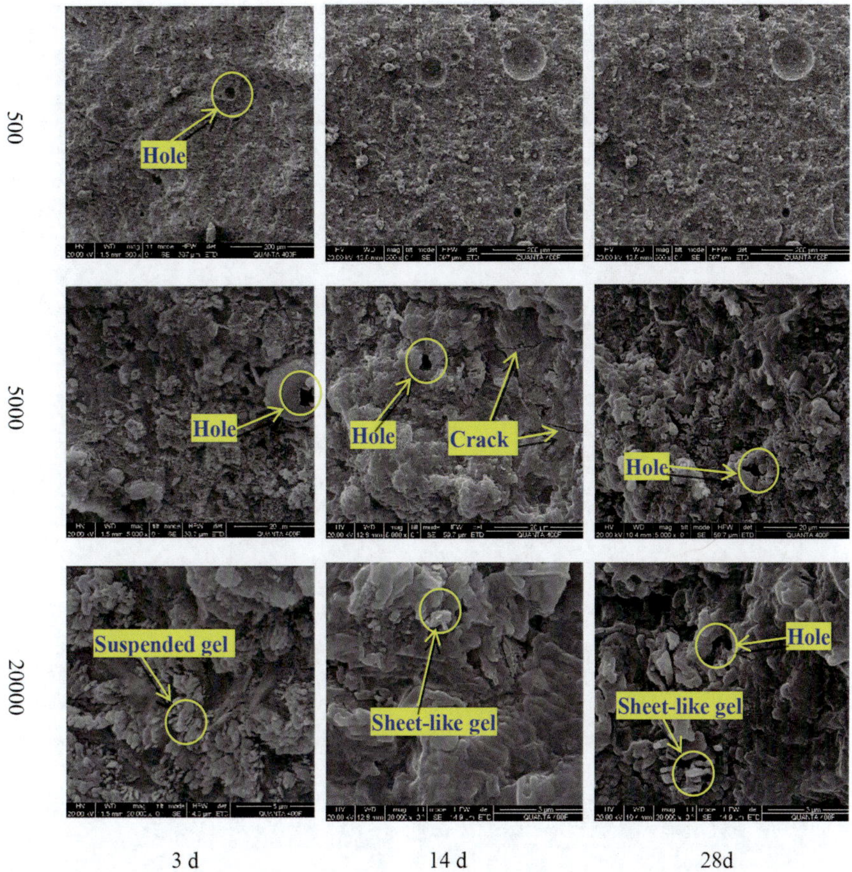

Fig. 3.9 Microscopic morphology of specimens at different curing ages

water-to-binder ratio increases, the microstructure of the specimens becomes looser, with an increase in pores and microcracks. The size of the pores between the flocculent gels also becomes larger, which aligns with the conclusion drawn from the MIP test. At a water-to-binder ratio of 0.55, needle-like ettringite and plate-like sodium sulfate can be observed among the hydration products, consistent with the analysis results of the XRD patterns. These microstructural observations are of critical significance for understanding the performance degradation mechanism of EFGM at higher water-to-binder ratios. The increased porosity and coarsened pore structure directly account for the reduction in compressive strength observed in Fig. 3.7, while the formation of needle-like ettringite and plate-like sodium sulfate at a water-to-binder ratio of 0.55 further confirms the detrimental effects of excessive free water on the stability of hydration products. From an engineering perspective, these findings underscore the necessity of strictly controlling the water-to-binder ratio (preferably ≤ 0.5) in practical grouting operations to maintain a dense microstructure and ensure adequate mechanical performance.

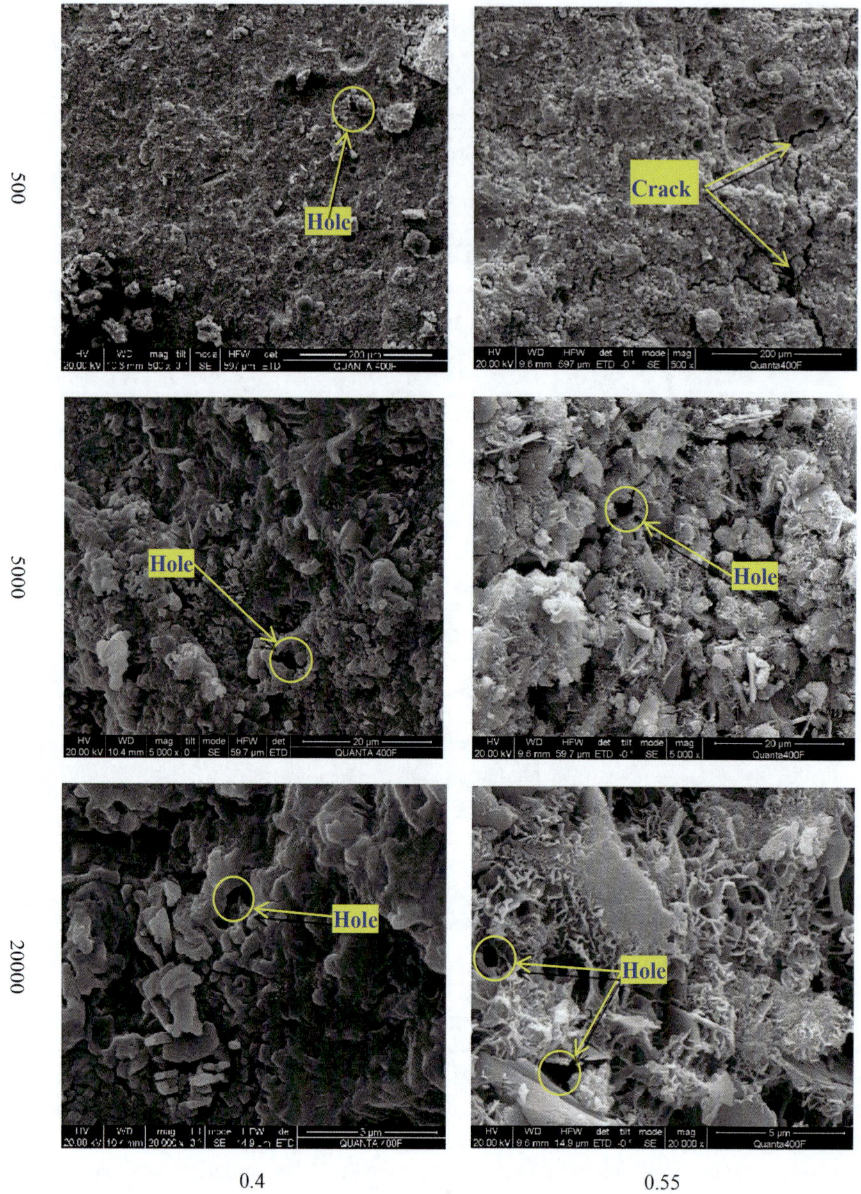

Fig. 3.10 Microscopic morphology of specimens with different water-cement ratios

3.2.3.4 MIP

The relevant microstructural parameters of specimens with different EMR contents, obtained from MIP tests, are presented in Table 3.1. As the EMR content increases, the porosity of the specimens increases. The trends for average pore diameter and apparent density first decrease and then increase, while the permeability first increases and then decreases.

The influence of different EMR contents on the pore structure of the specimens is shown in Fig. 3.11, with Fig. 3.11a presenting the cumulative distribution curve and Fig. 3.11b the incremental distribution curve. As the EMR content increases, the cumulative pore volume distribution of the specimens shows an upward shifting trend. Within the pore size range of 5–100 nm, the peak value corresponding to the incremental pore volume gradually rises with increasing EMR content.

According to relevant literature, pores are classified based on their diameters into harmless pores (<20 nm), less harmful pores (20–50 nm), harmful pores (50–200 nm), and multiharmful pores (>200 nm). The influence of different EMR contents on the pore size distribution of the specimens is shown in Fig. 3.12. As the EMR content increases, the porosity of the specimens exhibits an upward trend. Specifically, the proportion of harmless pores also increases; the proportion of less harmful pores first rises and then decreases; the proportion of harmful pores decreases; and the proportion of multiharmful pores first declines and subsequently increases.

The relevant parameters of specimens with different water-to-binder ratios, obtained from MIP tests, are listed in Table 3.2. As the water-to-binder ratio increases,

Table 3.1 Parameters of samples with different manganese waste content

Manganese slag content/%	Water-cement ratio	Average aperture/nm	Apparent density/(g/mL)	Porosity /%	Permeability / md
10	0.4	43.84	2.3276	33.2384	1131.57
30	0.4	32.86	2.0332	34.8847	2002.46
50	0.4	33.68	2.31	38.6717	279.2133

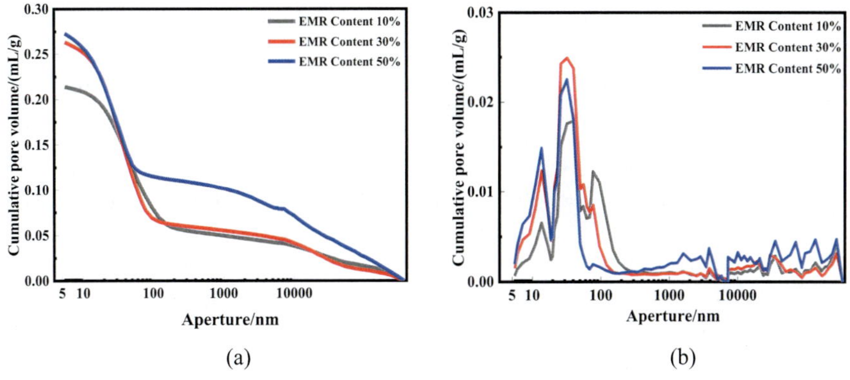

Fig. 3.11 The influence of different EMR contents on the pore structure of the specimen

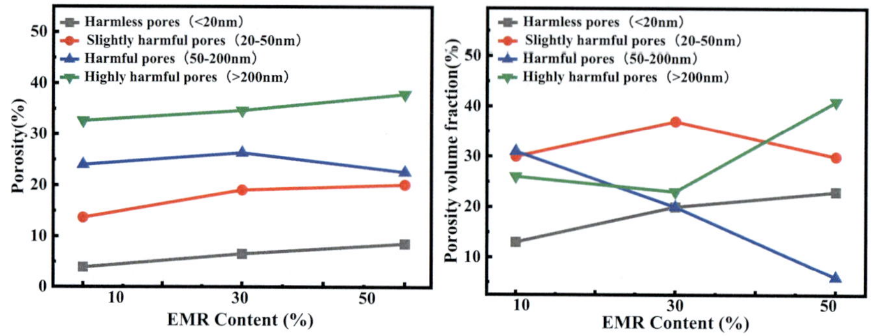

(a) (b)

Fig. 3.12 The influence of different EMR concentrations on the pore distribution of the specimens

Table 3.2 Parameters of Samples with Different Water-Cement Ratios

Manganese slag content/%	Water-cement ratio	Average aperture/nm	Apparent density/ (g/mL)	Porosity /%	Permeability /md
30	0.4	32.86	2.0332	34.8847	2002.46
30	0.55	71.46	2.3929	43.4493	697.6951

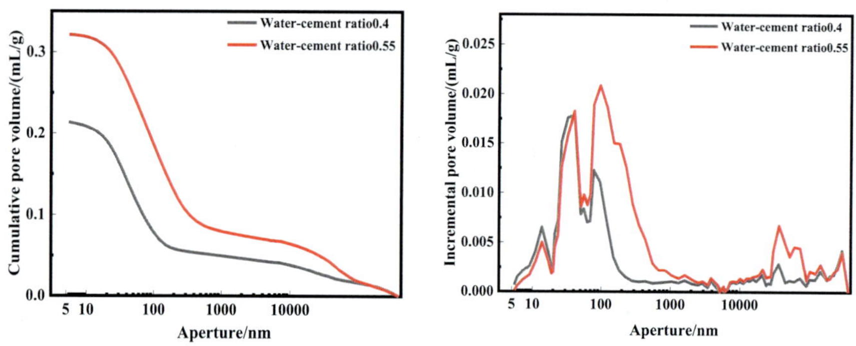

(a) Cumulative pore volume (b) Incremental pore volume

Fig. 3.13 The influence of different water-cement ratios on the pore structure of the specimens. (**a**) Cumulative pore volume. (**b**) Incremental pore volume

the porosity of the specimens rises significantly. The average pore diameter and apparent density also increase, whereas the permeability displays a decreasing trend.

The influence of different water-to-binder ratios on the pore structure of the specimens is presented in Fig. 3.13. As the water-to-binder ratio increases, the cumulative pore volume of the specimens also increases. Within the range of 70–1000 nm, the incremental pore volume within the same pore size range rises with a higher

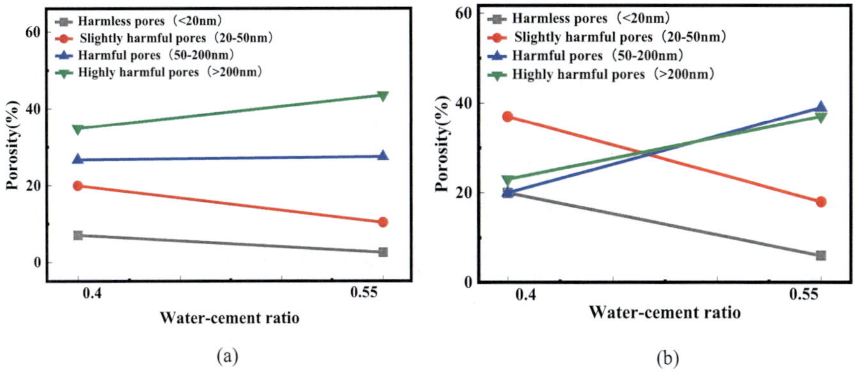

Fig. 3.14 The influence of different water-cement ratios on the pore distribution of the specimens

water-to-binder ratio. This indicates that, for pores of similar size, the number of pores increases within this specific range, thereby contributing to a higher overall porosity. In contrast, within the 5–70 nm range, the incremental pore volume curves for different water-to-binder ratios largely overlap, suggesting that variations in the water-to-binder ratio do not significantly affect the incremental pore volume for pores of equivalent size in this region. Overall, the increase in specimen porosity with a higher water-to-binder ratio is primarily attributed to the increase in the number of pores within the 70–1000 nm range.

The influence of different water-to-binder ratios on the pore size distribution of the specimens is shown in Fig. 3.14. As the water-to-binder ratio increases, the porosity of the specimens exhibits an increasing trend. Specifically, the proportions of both harmless pores and less harmful pores decrease, while the proportions of harmful pores and multiharmful pores increase.

3.3 Preparation of ESGM: An EMR-Based Grouting Material Incorporating GGBS and Steel Slag

3.3.1 Fresh Properties of the Grout

3.3.1.1 Flowability

Figure 3.15 illustrates the effects of EMR content and water-to-binder ratio on the flowability of the ESGM grout. As shown in the figure, an increase in EMR content, accompanied by a decrease in SS content, leads to a reduction in flowability. When the EMR content rises from 10% to 40%, the flowability of ESGM decreases from 241 mm to 195 mm. The decrease in flowability is relatively moderate when the

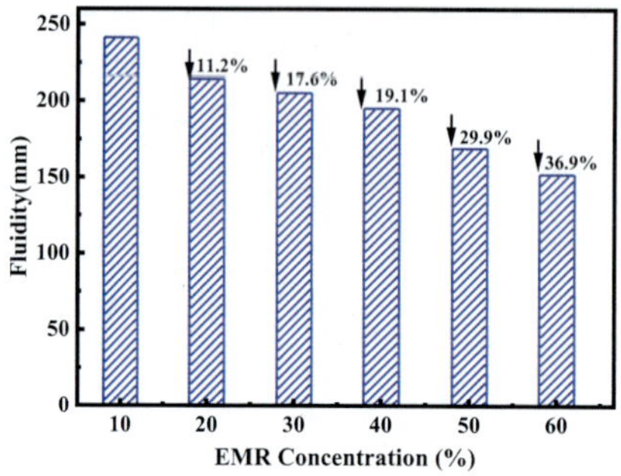

(a) The influence of the EMR and SS ratio on the fluidity of the ESGM slurry

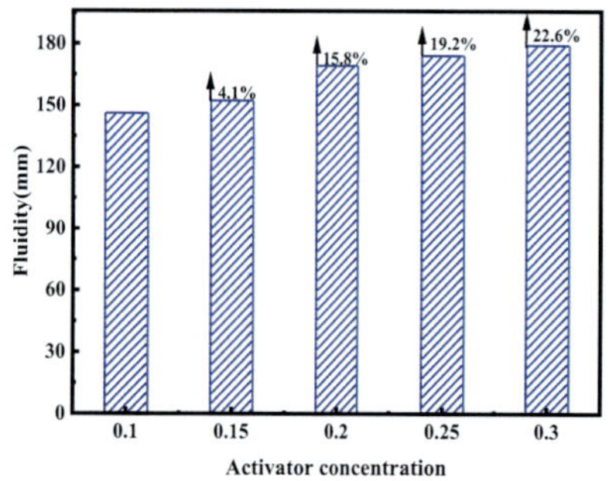

(b) The influence of the concentration of the activator on the fluidity of the ESGM slurry

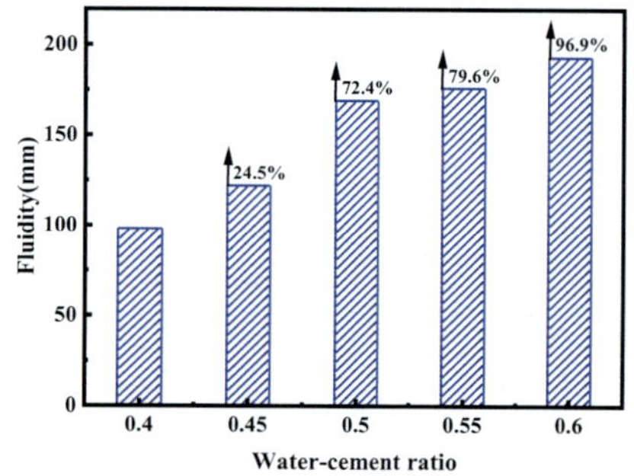

(c) The influence of water-cement ratio on the fluidity of ESGM slurry

Figure 3.19 Flowability of ESGM slurry

Fig. 3.15 Flowability of ESGM slurry. (**a**) The influence of the EMR and SS ratio on the fluidity of the ESGM slurry. (**b**) The influence of the concentration of the activator on the fluidity of the ESGM slurry. (**c**) The influence of water-cement ratio on the fluidity of ESGM slurry

EMR content is below 40%; however, a more pronounced reduction is observed when the EMR content exceeds 40%. Specifically, at an EMR content of 50%, the flowability is 169 mm, representing a decrease of 29.9% compared to that at 10% EMR content. At 60% EMR content, the flowability drops to 152 mm, corresponding to a reduction of 36.9% relative to the 10% EMR case.

Figure 3.15b shows the influence of alkali-activator concentration on the flowability of the ESGM grout at a fixed EMR content of 50% and a water-to-binder ratio of 0.5. The effect of activator concentration on flowability is relatively limited, with flowability values ranging between 98 mm and 174 mm. The flowability increases as the activator concentration rises. At activator concentrations of 0.1, 0.15, 0.2, 0.25, and 0.3, the corresponding flowability values are 146 mm, 152 mm, 169 mm, 174 mm, and 179 mm, respectively. When the activator concentration is 0.3, the flowability increases by 22.6% compared to that at 0.1.

Figure 3.15c presents the effect of water-to-binder ratio on flowability at an EMR content of 50% and an activator concentration of 0.2. The water-to-binder ratio has a significant impact on flowability, with flowability increasing markedly as the water-to-binder ratio is raised, showing a clear positive correlation. At water-to-binder ratios of 0.4, 0.45, 0.5, 0.55, and 0.6, the flowability values are 98 mm, 122 mm, 169 mm, 176 mm, and 193 mm, respectively. The flowability at a water-to-binder ratio of 0.6 is 96.9% higher than that at 0.4.

The influence of EMR content on grout flowability can be attributed to two main aspects. First, it relates to the physical properties of the raw materials. Differences in the physical characteristics of various powders affect the flowability of the grout. EMR contains a substantial amount of hydrated calcium sulfate; after calcination, it consists mainly of anhydrous calcium sulfate, giving it strong water-absorbing properties. In contrast, SS has low reactivity, poor early-age reaction kinetics, and requires less water. Therefore, an increase in EMR content and a corresponding decrease in SS content reduce the amount of free water in the grout, thereby decreasing its flowability. Conversely, a higher SS content increases flowability, which is consistent with previous studies. Second, chemical reactions between the raw materials generate cementitious products. After thermal activation, the reactivity of EMR is enhanced. An increase in EMR content raises the amount of reactive components in the grout. Moreover, the calcium sulfate present in EMR promotes the rapid formation of ettringite, which reduces grout flowability.

An increase in activator concentration strengthens the electrostatic repulsion between particles, affecting the microstructure of the grout. This leads to more uniform particle dispersion, smaller flocculated structures, and a narrower particle size distribution. The microscopic morphology shifts from clustered to chain-like structures, reducing the coverage of free water by flocculated networks and enhancing the dispersing effect of free water on particles, thereby increasing flowability.

An increase in the water-to-binder ratio raises the amount of free water in the grout system, facilitating flow. The additional water increases the spacing between raw material particles, hindering chemical reactions and reducing the formation of cementitious products, which consequently enhances grout flowability.

3.3.1.2 Initial and Final Setting Times

Figure 3.16 presents the setting times of the ESGM grout under different EMR contents, alkali-activator concentrations, and water-to-binder ratios. The effect of varying EMR content on the setting times is shown in Fig. 3.16a. It can be observed that the setting times decrease as the EMR content increases, which corresponds to a decrease in SS content. When the EMR content varies from 10% to 60%, the initial setting time is shortened from 171 min to 97 min (a reduction of 43.3%), and the final setting time is shortened from 186 min to 123 min (a reduction of 33.9%). SS possesses the ability to delay setting and reduce heat of hydration. As indicated in Table 2.1, SS has a CaO content of 26.85%, reflecting its low reactivity. Therefore, when the SS content is high and the EMR content is low, fewer cementitious products are formed, resulting in longer grout setting times. In contrast, calcined EMR exhibits significantly enhanced reactivity and contains calcium sulfate, which

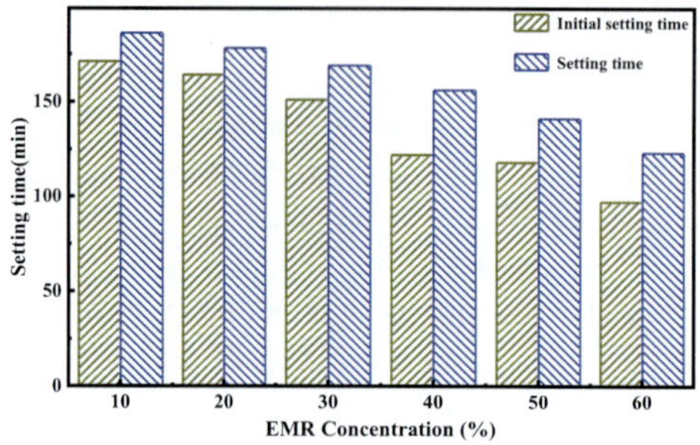

(a) The influence of the EMR and SS ratio on the setting time of ESGM slurry

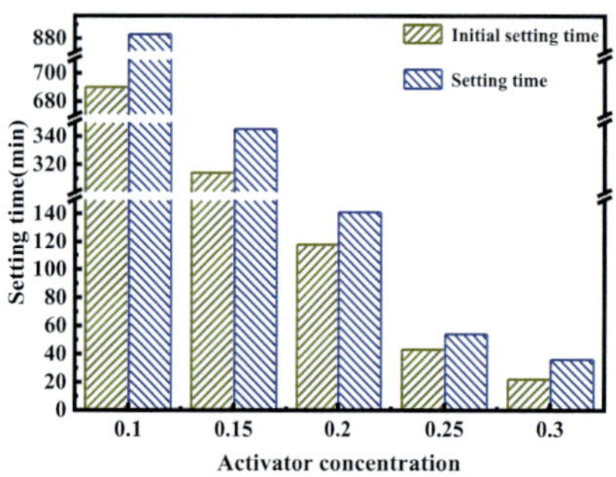

(b) The influence of the concentration of the activator on the setting time of ESGM slurry

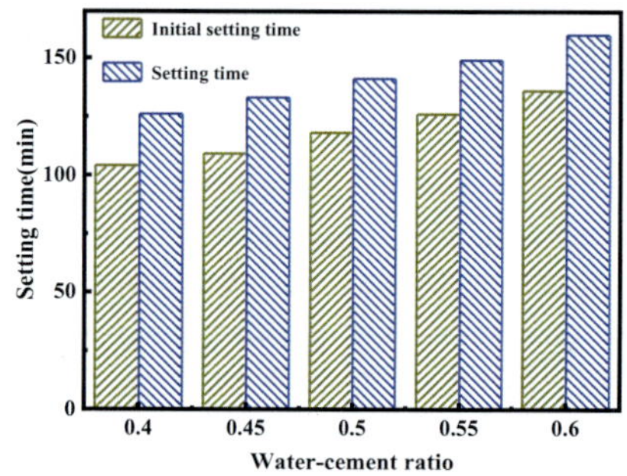

(c) The influence of water-cement ratio on the setting time of ESGM slurry

Figure 3.20 ESGM slurry setting time

Fig. 3.16 ESGM slurry setting time. (**a**) The influence of the EMR and SS ratio on the setting time of ESGM slurry. (**b**) The influence of the concentration of the activator on the setting time of ESGM slurry. (**c**) The influence of water-cement ratio on the setting time of ESGM slurry

absorbs a portion of the free water in the grout. Consequently, when SS content is low and EMR content is high, the reactive components in the alkali-activated system are sufficiently stimulated and dissolved to form C-(A)-S-H gels. Some of the C-(A)-S-H gels further react with $CaSO_4$ to form ettringite (AFt) or monosulfoaluminate (AFm). During gel formation, the consumption of calcium ions promotes the continuous dissolution of raw materials, leading to the generation of more gel products and thereby shortening the grout setting time. Additionally, sulfate can activate the reactivity of SS and GGBS. Hence, an increase in EMR content implies a higher calcium sulfate content in the system, which activates SS and GGBS to some extent, accelerates gel formation, and consequently reduces the grout setting time.

Figure 3.16b illustrates the influence of activator concentration on the setting times of the ESGM grout. The results indicate that the setting times decrease as the activator concentration increases. At an activator concentration of 0.1, the initial and final setting times are 690 min and 883 min, respectively. When the activator concentration is 0.2, the initial and final setting times decrease significantly to 118 min and 141 min, respectively. Below an activator concentration of 0.2, the setting times increase markedly. At activator concentrations ≥0.2, the variation in setting times becomes relatively minor; for instance, at a concentration of 0.3, the initial and final setting times are 22 min and 36 min, respectively. An increase in activator concentration raises the alkali content in the grout. The elevated alkali content accelerates the breakdown of chemical bonds in the raw materials and promotes their dissolution, enabling rapid formation of gel products and consequently reducing the setting time. However, when used as a grouting material, sufficient time is often required for pumping operations. Excessively short setting times can hinder the pumping process and affect construction progress and effectiveness; therefore, the activator concentration should not be excessively high.

The effect of water-to-binder ratio on the setting times of the ESGM grout is shown in Fig. 3.16c. The setting times increase with an increase in the water-to-binder ratio. As the water-to-binder ratio increases from 0.4 to 0.6, the corresponding initial setting times are 104 min, 109 min, 118 min, 126 min, and 136 min, and the final setting times are 126 min, 133 min, 141 min, 149 min, and 160 min, respectively. Compared to the effect of activator concentration, the influence of the water-to-binder ratio on setting times is relatively moderate. An increase in the water-to-binder ratio elevates the amount of free water in the grout system. The excess water cannot be fully consumed during the reaction and simultaneously disperses the solid particles, increasing the inter-particle distance. This further reduces the reaction rate, slows the formation of cementitious products, and weakens the bonding between gels. Consequently, more time is required for the grout to form a fundamental cementitious skeleton, resulting in prolonged setting times, which aligns with observations reported in previous studies.

3.3.2 Strength Characteristics of the Grouting Material

3.3.2.1 Effect of EMR Content on Compressive Strength

Figure 3.17 illustrates the effect of different EMR contents on the compressive strength of ESGM. As shown in Fig. 3.17a, the strength of ESGM increases with prolonged curing age and exhibits an initial enhancement followed by a subsequent decline as the EMR content increases. The highest compressive strength was achieved at an EMR content of 50% and an SS content of 20%, with 3-day, 14-day, and 28-day compressive strengths reaching 12.6 MPa, 14.2 MPa, and 15.4 MPa, respectively. Beyond an EMR content of 50%, the strength of ESGM decreased. At an EMR content of 10% and an SS content of 50%, the corresponding 3-day, 14-day, and 28-day compressive strengths were 2.3 MPa, 4.4 MPa, and 5.5 MPa, respectively.

When the EMR contents were 20%, 30%, 40%, 50%, and 60%, the 3-day compressive strengths increased by 65.2%, 113.1%, 243.5%, 447.8%, and 365.2%, respectively, compared to the sample with 10% EMR content. Similarly, the 28-day compressive strengths showed increases of 10.5%, 44.8%, 67.6%, 143.3%, and

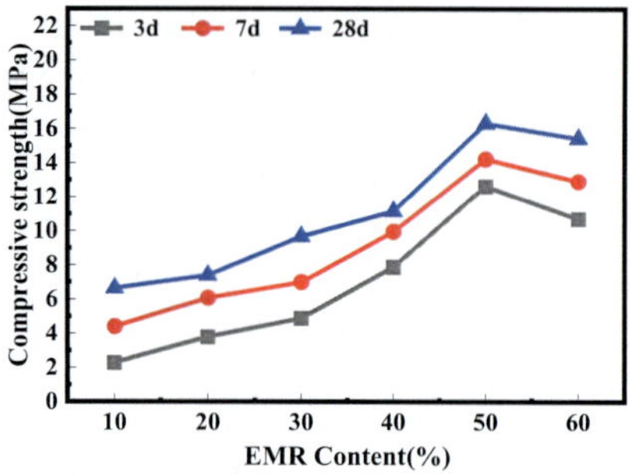

(a) The compressive strength of ESGM under different EMR and SS ratios

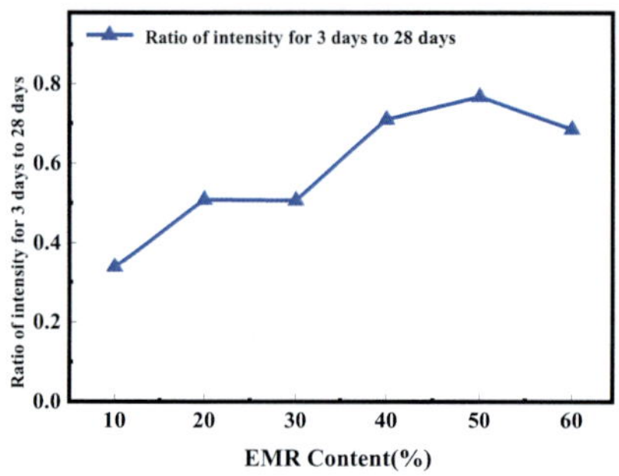

(b) Ratio of intensity for 3 days to 28 days

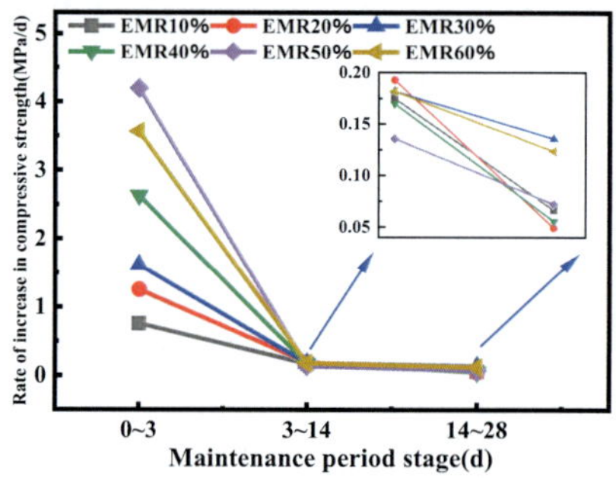

(c) The growth rate of compressive strength in different age groups

Fig. 3.17 The influence of the ratio of EMR to SS on the compressive strength of ESGM. (**a**) The compressive strength of ESGM under different EMR and SS ratios. (**b**) Ratio of intensity for 3 days to 28 days. (**c**) The growth rate of compressive strength in different age groups

129.9%, respectively. The percentage growth in 3-day compressive strength was more pronounced than that in 28-day strength, indicating that the incorporation of EMR contributes more significantly to early-age strength development—a critical factor for grouting operations. The optimal mix proportion for compressive strength was identified as 50% EMR, 20% SS, and 30% GGBS, under which the 28-day compressive strength reached 15.4 MPa.

Figure 3.17c shows the average compressive strength growth rates across different curing periods under varying EMR contents, providing a clearer visualization of the influence of EMR content on the compressive strength of the specimens at different ages. During the 0–3 day curing period, the specimens exhibited the highest average compressive strength growth rate, which continuously increased with rising EMR content, reaching its maximum at 50% EMR. However, a decrease in the growth rate was observed when the EMR content reached 60%. In the subsequent curing intervals of 3–14 days and 14–28 days, the contribution of EMR content to the strength growth rate diminished significantly. The compressive strength growth rates of specimens with different EMR contents became largely comparable, showing no substantial variation. This indicates that increasing the EMR content primarily enhances early-age strength (0–3 days). The main phase of strength development occurred within the 0–14 day curing period, while the strength improvement during the 14–28 day period was relatively limited.

The observed phenomena can be attributed to the high content of $CaSO_4$ in EMR. As the EMR content increases, the $CaSO_4$ content in the grout rises correspondingly. Under alkaline conditions, $CaSO_4$ participates in chemical reactions, promoting the formation of more ettringite. As a strength-enhancing phase, ettringite contributes to the increased strength of ESGM. The formation of geopolymers also positively contributes to the strength of ESGM. After calcination, the low-crystallinity $CaSO_4{\cdot}2H_2O$ and $CaSO_4{\cdot}0.5H_2O$ in EMR undergo thermal decomposition to form CaO. Additionally, part of the originally inert SiO_2 and Al_2O_3 transforms into reactive SiO_2 and Al_2O_3, substantially enhancing the reactivity of EMR.

When the EMR content reaches 60%, several factors may lead to a reduction in ESGM strength. On one hand, $CaSO_4$ exhibits high water absorption. The hydrated calcium sulfate and gypsum formed during the reaction contain crystalline water, which can cause volume expansion in ESGM. This expansion may introduce internal microcracks, compromising the structural integrity of the material and consequently reducing its strength. On the other hand, the particle size of the raw materials also plays an important role in strength development. The particle size and specific surface area of SS are crucial for the development of strength in cementitious products. Generally, a larger particle size of raw materials slows down their dissolution rate in the system. In this study, the SS used had a relatively coarse particle size, which contributed to the lower compressive strength observed in specimens with high SS content. Nevertheless, a portion of the SS may act as fine aggregate in the specimens, providing some degree of strength support.

3.3.2.2 Effect of Activator Concentration

Figure 3.18 illustrates the effect of different activator concentrations on the compressive strength of ESGM. As shown in Fig. 3.18a, the compressive strength of ESGM initially increases and then decreases with increasing activator concentration. The maximum compressive strength is achieved at an activator concentration of 0.2. When the activator concentration is 0.1, the compressive strengths at 3, 14, and 28 days are 2.3 MPa, 6.2 MPa, and 7.8 MPa, respectively. At activator concentrations of 0.15, 0.2, 0.25, and 0.3, the 3-day compressive strengths increase by 130.4%, 447.8%, 282.6%, and 217.4%, respectively, compared to the strength at an activator concentration of 0.1. Similarly, the corresponding 28-day compressive strengths increase by 13.5%, 83.1%, 58.4%, and 42.7%. These results indicate that the growth in 3-day strength is significantly more pronounced than that in 28-day strength, suggesting that a higher activator concentration is not necessarily better. An excessively high concentration can adversely affect the strength, while an appropriate concentration is beneficial for enhancing early-age strength and is more

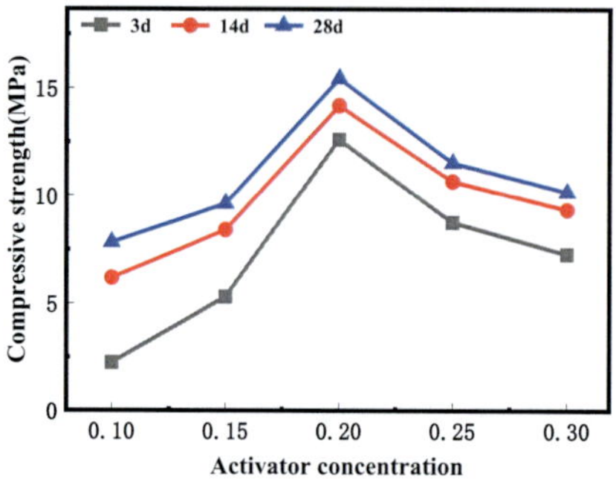

(a) Compressive strength of ESGM under different concentrations of excitation agents

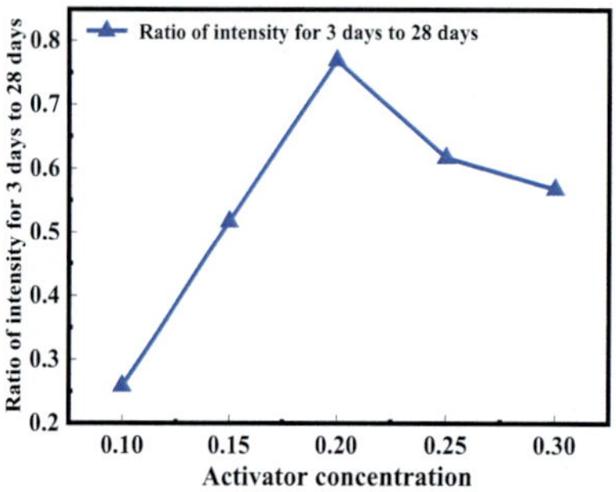

(b) Ratio of intensity for 3 days to 28 days

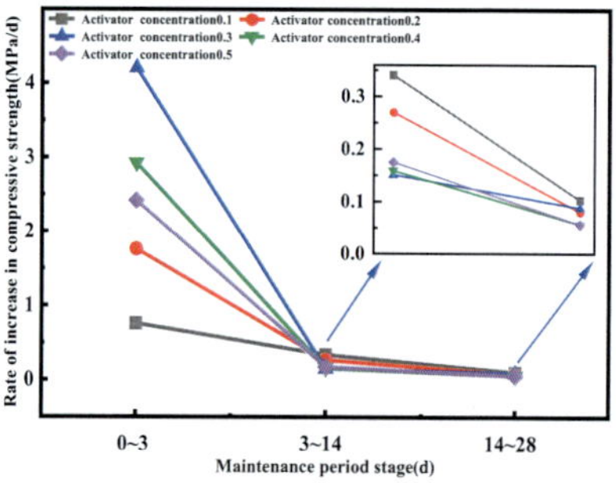

(c) The growth rate of compressive strength in different age groups

Fig. 3.18 The influence of activator concentration on the compressive strength of ESGM. (**a**) Compressive strength of ESGM under different concentrations of excitation agents. (**b**) Ratio of intensity for 3 days to 28 days. (**c**) The growth rate of compressive strength in different age groups

economical. Based on comprehensive analysis and comparison in this study, the optimal activator concentration is determined to be 0.2.

Figure 3.18c presents the average compressive strength growth rates within different curing periods under various activator concentrations. It can be observed that the average strength growth rate during the 0–3 day period is the fastest among all age ranges and is closely related to the activator concentration. Specifically, the growth rate initially increases and then decreases with increasing activator concentration. In the subsequent periods of 3–14 days and 14–28 days, the contribution of activator concentration to the strength growth rate diminishes significantly. The compressive strength growth rates under different activator concentrations become nearly identical, indicating that early-age strength development mainly occurs within 0–14 days, while the strength improvement during 14–28 days is relatively limited.

When the activator concentration is too low, the alkali content in the grout system is insufficient to effectively activate the reactive components in the raw materials. As a result, fewer cementitious products are formed, which is unfavorable for strength development. Conversely, when the activator concentration is excessively high, the alkalinity of the system becomes too elevated, causing the reactive components to be rapidly activated and form gels that coat the surface of raw material particles. This coating hinders further alkali activation, preventing some raw materials from being effectively activated and thereby negatively affecting the strength, which fails to reach the desired level.

3.3.2.3 Effect of Water-to-Binder Ratio

Figure 3.19 shows the compressive strength of ESGM under different water-to-binder ratios. As illustrated in Fig. 3.19a, the compressive strength of ESGM exhibits a decreasing trend with an increase in the water-to-binder ratio, reaching its maximum value at a water-to-binder ratio of 0.4. Specifically, at a water-to-binder ratio of 0.4, the 3-day, 14-day, and 28-day compressive strengths of the specimen

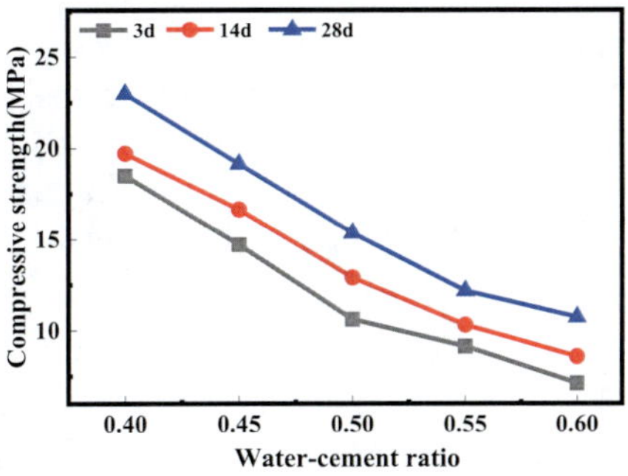

(a) Compressive strength of ESGM under different water-cement ratios

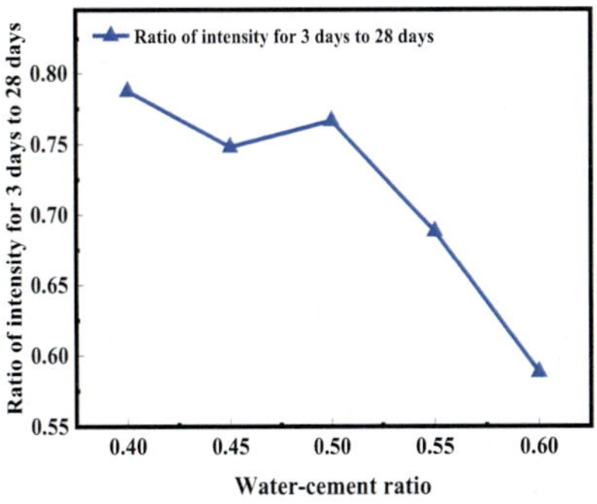

(b) Ratio of intensity for 3 days to 28 days

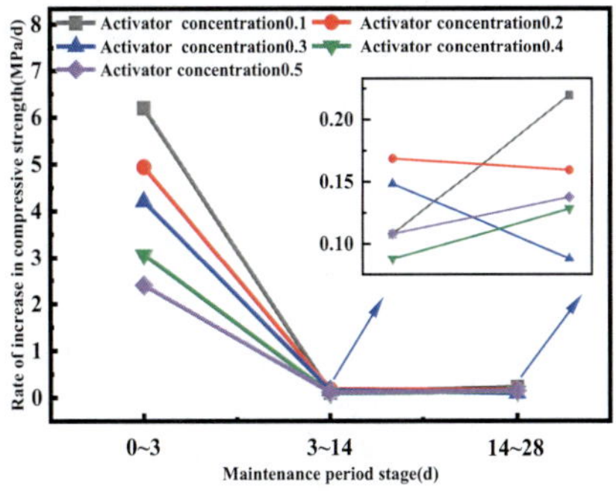

(c) The growth rate of compressive strength in different age groups

Fig. 3.19 The influence of water-cement ratio on the compressive strength of ESGM. (**a**) Compressive strength of ESGM under different water-cement ratios. (**b**) Ratio of intensity for 3 days to 28 days. (**c**) The growth rate of compressive strength in different age groups are 18.6 MPa, 19.8 MPa, and 23.0 MPa, respectively. When the water-to-binder ratios are 0.45, 0.5, 0.55, and 0.6, the 3-day compressive strengths decrease by 20.4%, 32.3%, 50.5%, and 59.7%, respectively, compared to that at a water-to-binder ratio of 0.4. Correspondingly, the 28-day compressive strengths decrease by 15.4%, 30.3%, 42.7%, and 47.9%, respectively.

The water-to-binder ratio exerts a significant influence on the compressive strength of the specimens. This is because an increase in the water-to-binder ratio leads to a larger inter-particle distance among the cementitious materials, making it more difficult to form an effectively continuous skeletal matrix. Additionally, the higher water content in the grout results in increased porosity of the specimen, thereby reducing its compressive strength.

The average compressive strength growth rates of specimens under different water-to-binder ratios across various curing periods are shown in Fig. 3.19c. It can be observed that the average compressive strength growth rate during the 0–3 day period is the highest, and this rate decreases as the water-to-binder ratio increases. In the subsequent periods of 3–14 days and 14–28 days, the strength growth rates decrease significantly and show little variation among different water-to-binder ratios. This indicates that the hydration reaction of the grout system mainly occurs in the early stage, and the strength enhancement of the specimens primarily takes place within the 0–3 day period. By 28 days, the hydration reaction is essentially complete, and further contribution to strength improvement becomes limited. Therefore, the strength development process is critical in the early stage and gradually stabilizes in later periods.

3.3.3 Microstructural Characteristics of the Grouting Material

3.3.3.1 XRD

Figure 3.20 presents the XRD patterns of ESGM specimens with a curing age of 28 days under different EMR contents and activator concentrations. The crystalline minerals identified in the XRD patterns mainly include quartz (SiO_2, PDF#85–0335), ettringite ($Ca_6Al_2(SO_4)_3(OH)_{12}{\cdot}26H_2O$, PDF#02–0059), manganese dioxide (MnO_2, PDF#72–1982), RO phase (CaO-FeO-MnO-MgO solid solution, PDF#77–2379, PDF#77–2355), manganite hydrate ($H_2Mn_8O_{16}{\cdot}2.4H_2O$, PDF#48–0876), calcite ($CaCO_3$, PDF#03–0593), and muscovite ($KAl_2Si_3AlO_{10}(OH)_2$, PDF#07–0032).

In Fig. 3.20a, compared with the XRD patterns of the raw materials (Figs. 2.2 and 2.5), it can be seen that quartz and muscovite are primarily derived from EMR, with a portion of the activated SiO_2 participating in the geopolymerization reaction. The RO phase mainly originates from SS, and its diffraction peaks are more intense at higher SS contents. The diffraction peaks of hydrated calcium sulfate ($CaSO_4{\cdot}2H_2O$) and hemihydrate gypsum ($CaSO_4{\cdot}0.5H_2O$) from EMR, as well as C_2S and C_3S from SS, disappear due to hydration, leading to the formation of ettringite, which has also been observed in previous literature. MnO_2 is formed

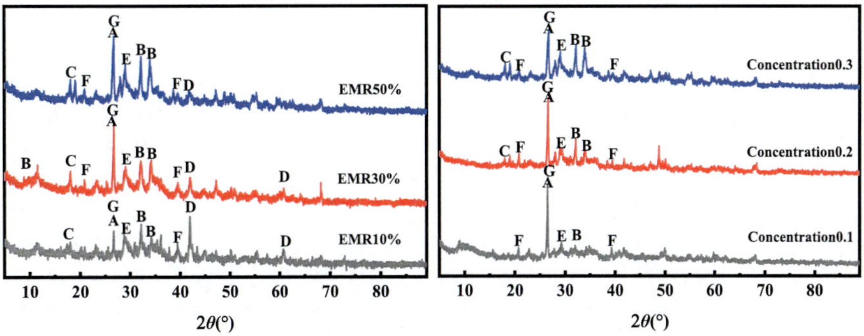

(a) EMR and SS ratio (b) Activator concentration

Fig. 3.20 XRD spectrum of ESGM sample after 28 days. (A: Quartz (SiO2); B: Calcium feldspar (Ca6Al2(SO4)3(OH)12·26H2O); C: Manganese dioxide (MnO2); D: RO phase (CaO-FeO-MnO-MgO solid solution); E: Manganese hydroxide hydrate (H2Mn8O16·2.4H2O); F: Calcite (CaCO3); G: Pyroxene (KAl2Si3AlO10(OH)2)

through the oxidation of Mn^{2+} during calcination, while manganite hydrate is generated in the alkaline environment through the combination of MnO_2, Mn^{2+}, and OH^-. Calcite is formed by the reaction of CaO, water, and carbon dioxide during the curing process. From Fig. 3.20b, it can be observed that higher activator concentrations result in stronger diffraction peaks for ettringite and manganite hydrate, indicating that an increased alkali-activator concentration favors their formation.

The diffraction features in the range of 20°–40° (2θ) correspond to the amorphous gel products of C-(A)-S-H and N-A-S-H formed during the alkali-activation reaction. Such a broad hump is also observed in Fig. 3.20, confirming the formation of C-(A)-S-H and N-A-S-H composite gels, which contribute to the strength of the specimens. The variation in the area of the amorphous hump reflects the quantity of the cementitious products formed—a larger hump area indicates a greater amount of amorphous gels. In Fig. 3.20a, it can be seen that as the EMR content increases, the hump area within 20°–40° (2θ) also becomes larger, suggesting the formation of more amorphous C-(A)-S-H and N-A-S-H gels, which corresponds to higher specimen strength. In Fig. 3.20b, the hump area increases with rising activator concentration, indicating that a higher activator concentration promotes the formation of more amorphous C-(A)-S-H and N-A-S-H gels. However, an increased activator concentration also enhances the formation of ettringite. While ettringite contributes to early-age strength in hardened specimens, it can also cause volumetric expansion, leading to more microcracks within the specimen and thereby reducing its strength. Consequently, specimens with an activator concentration of 0.2 exhibit higher strength than those with a concentration of 0.3.

3.3.3.2 FTIR

The FTIR spectra of ESGM specimens at 28 days are presented in Fig. 3.21. The vibration absorptions at 3641 cm^{-1}, 3423–3447 cm^{-1}, and 1641–1648 cm^{-1} can be attributed to O–H groups, indicating the presence of water. In Fig. 3.21b, the absorption peak at 3641 cm^{-1} observed at an activator concentration of 0.3 is likely caused by the O–H vibration from NaOH.

The vibration absorptions around 1429 cm^{-1}, 1485 cm^{-1}, 1486 cm^{-1}, and 1488 cm^{-1} correspond to O–C–O groups, suggesting the formation of carbonates. As shown in Fig. 3.21b, two absorption peaks appear in this region at an activator concentration of 0.3, which, combined with the XRD results, may be assigned to calcite and sodium carbonate formed from the carbonation of excess activator.

The absorptions near 1103 cm^{-1}, 1115 cm^{-1}, 1117 cm^{-1}, 1119 cm^{-1}, as well as around 798 cm^{-1} and 875 cm^{-1}, are assigned to Si–O–Si(Al) groups, confirming the formation of C-(A)-S-H and N-A-S-H gels. With higher EMR content (Fig. 3.21a), the intensity of these absorption peaks increases, indicating a greater amount of C-(A)-S-H and N-A-S-H gels formed. Conversely, when the activator concentration is too low (e.g., 0.1, Fig. 3.21b), the peak intensity in this region is weaker, reflecting lower gel formation.

The vibration bands at 954 cm^{-1}, 958 cm^{-1}, and 997 cm^{-1} are attributed to Mn–O groups, suggesting the presence of MnO_2 and manganese hydroxide hydrate. As the EMR content increases (Fig. 3.21a), the intensity of these absorptions strengthens, indicating higher MnO_2 content. With increasing activator concentration (Fig. 3.21b), the peaks become stronger and shift to higher wavenumbers, implying that a higher activator concentration promotes the formation of manganese hydroxide hydrate.

The small absorption bands near 639 cm^{-1} and 616 cm^{-1} are assigned to SO_4^{2-} groups, representing the formation of ettringite. An increase in EMR content

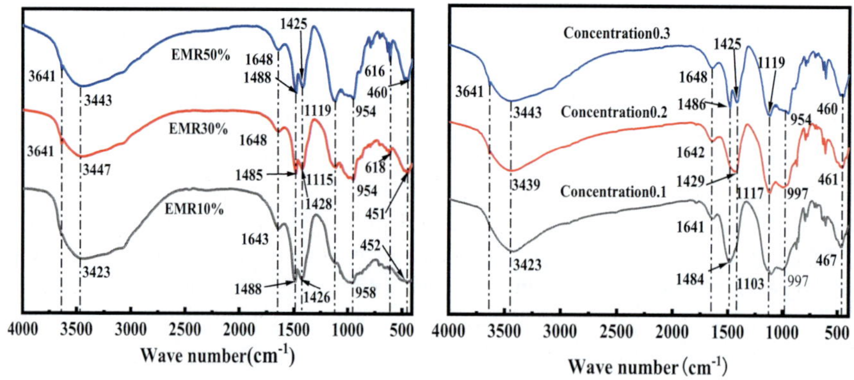

(a) EMR and SS ratio

(b) Activator concentration

Fig. 3.21 FTIR spectrum of ESGM sample after 28 days. (**a**) EMR and SS ratio. (**b**) Activator concentration enhances the ettringite absorption peak (Fig. 3.21a), indicating more ettringite formation.

The vibration absorptions at 460 cm^{-1}, 461 cm^{-1}, 467 cm^{-1}, 452 cm^{-1}, and 451 cm^{-1} are related to SiO_2, which originates from the raw materials. The strengthening of these peaks with higher EMR content (Fig. 3.21a) suggests that SiO_2 in the specimens mainly comes from EMR.

The FTIR analysis is consistent with the mineral phases identified in the XRD patterns.

3.3.3.3 SEM-EDS

Figure 3.22 shows the SEM images and EDS results of the specimen with 10% EMR and 60% SS at 28 days. The images reveal loose pores, cracks, acicular/rod-like gels, and amorphous gels, indicating a relatively loose structure with large pores, weak inter-particle connections, and poor integrity. Under alkali activation,

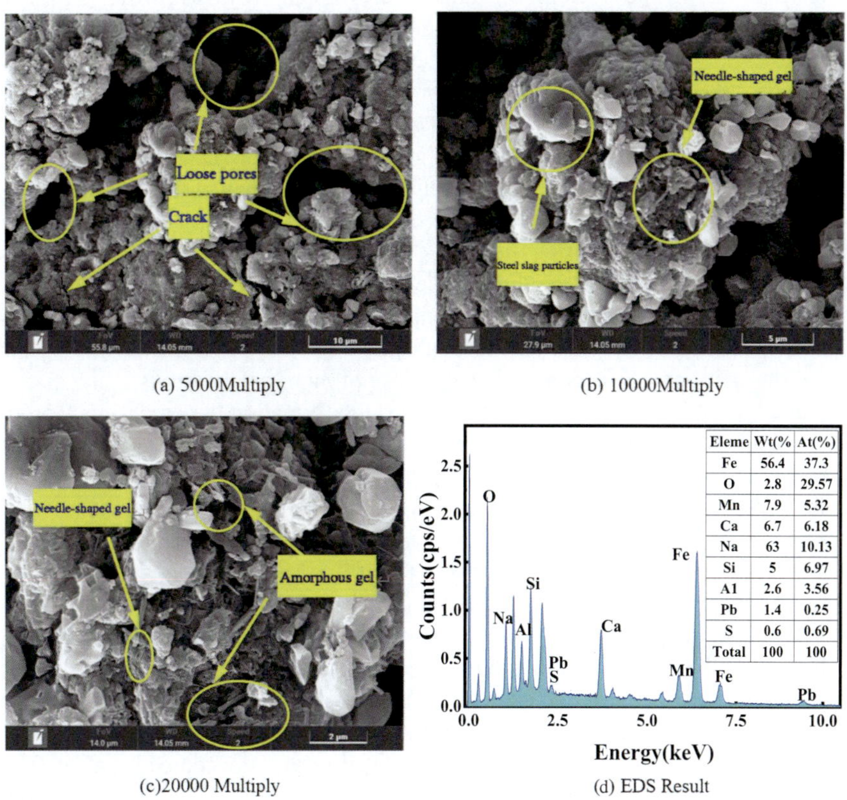

Fig. 3.22 SEM image and EDS results of EMR (10%) + SS (60%) sample after 28 days. (**a**) 5000 Multiply. (**b**) 10000 Multiply. (**c**) 20,000 Multiply. (**d**) EDS Result

EMR, SS, and GGBS dissolve to form cementitious products. While EMR and GGBS are largely consumed, the low reactivity of SS results in its limited consumption. Consequently, the amount of cementitious material formed is insufficient, leading to a loose structure with large pores and cracks between particles and weak bonding. This suggests that the acicular/rod-like and amorphous gels are inadequate to fill the pores and cracks, which macroscopically corresponds to lower strength. The cracks present in the matrix may be caused by stress concentration around large pores and shrinkage stress.

Figure 3.22d presents the EDS analysis results of the cementitious products. The main elements detected are Fe, Mn, O, Ca, Na, Si, and Al. The atomic ratios of Ca/Si, Al/Si, and Na/Si in the gel products are 0.89, 0.51, and 1.45, respectively. The high content of Fe, primarily derived from SS, indicates that a portion of SS participated in the hydration reaction and formed gels. The elements Mn and Pb mainly originate from EMR, showing that heavy metals are immobilized within the cementitious products. The presence of S suggests the formation of ettringite (acicular/rod-like gels). The high Na content indicates the occurrence of geopolymerization, generating typical geopolymeric cementitious material—N-A-S-H gel. The presence of Ca, Si, and Al elements confirms the formation of C-(A)-S-H gel, which coexists with the N-A-S-H gel, contributing to the strength of the specimen.

Figure 3.23 shows the SEM images and EDS results of the specimen with 30% EMR and 40% SS at 28 days. Loose pores, cracks, and amorphous network-like gels can be observed. Compared to the specimen with 10% EMR and 60% SS, the 30% EMR + 40% SS specimen forms a greater amount of cementitious material, resulting in a denser structure and relatively smaller pores. In Fig. 3.23a, amorphous network-like gels formed by interconnected cementitious materials can be observed, which were hardly seen in the 10% EMR + 60% SS specimen. These amorphous network-like gels fill the larger pores of the specimen, reducing pore size and porosity, thereby contributing to enhanced strength. Additionally, the formed amorphous gels are observed coating the surfaces of unhydrated SS particles, improving the overall integrity of the specimen. The presence of more gel materials in the matrix indicates that a higher EMR content promotes the formation of cementitious products. Calcined EMR exhibits higher reactivity than SS, which can be attributed to its increased content of CaO, reactive SiO_2, and Al_2O_3, leading to the generation of more C-(A)-S-H and N-A-S-H gels in the specimen.

Figure 3.23d presents the EDS analysis results of the cementitious products. The elemental composition is similar to that of the 10% EMR + 60% SS specimen. The atomic ratios of Ca/Si, Al/Si, and Na/Si in the gel products of the 30% EMR + 40% SS specimen are 1.13, 0.29, and 0.65, respectively. Compared to the 10% EMR + 60% SS specimen, the Ca/Si ratio increases, while the Al/Si and Na/Si ratios decrease. The EDS results indicate that the increase in EMR raises the Ca and Si content in the cementitious materials, promoting the formation of more C-(A)-S-H and N-A-S-H gels. The decrease in Fe is related to the reduced SS content. The sulfur content increases to 1.76%, suggesting greater ettringite formation. Furthermore, heavy metals were effectively immobilized within the cementitious products.

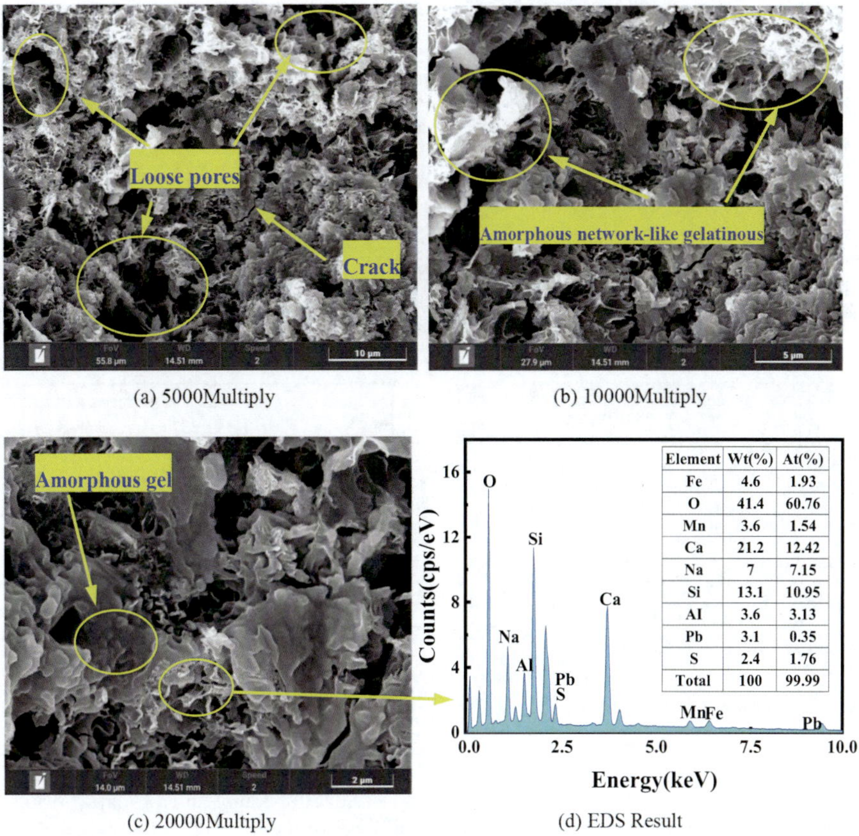

Fig. 3.23 SEM image and EDS results of EMR (30%) + SS (40%) sample at 28 days. (**a**) 5000 Multiply. (**b**) 10,000 Multiply. (**c**) 20,000 Multiply. (**d**) EDS result

Figure 3.24 presents the SEM images and EDS results of the specimen with 50% EMR and 20% SS at 28 days. Compared to the specimens with 10% EMR + 60% SS and 30% EMR + 40% SS, the 50% EMR + 20% SS specimen exhibits a significantly denser microstructure, better integrity, and noticeably reduced porosity. The cementitious materials form a relatively complete matrix, and the connections between the cementitious products are much more compact. Very few SS particles are visible in the matrix, and the amorphous gels are tightly packed, forming a dense matrix with minimal observable pores. This indicates that an increased EMR content is beneficial for promoting the formation of more cementitious substances in the matrix, thereby enhancing the macroscopic strength of the specimen. Additionally, minor microcracks are observed in the matrix, which are likely attributed to reaction-induced shrinkage.

Figure 3.24d shows the EDS analysis results of the cementitious products. The primary elemental composition is consistent with that of the 10% EMR + 60% SS and 30% EMR + 40% SS specimens. In the gel products of the 50% EMR + 20%

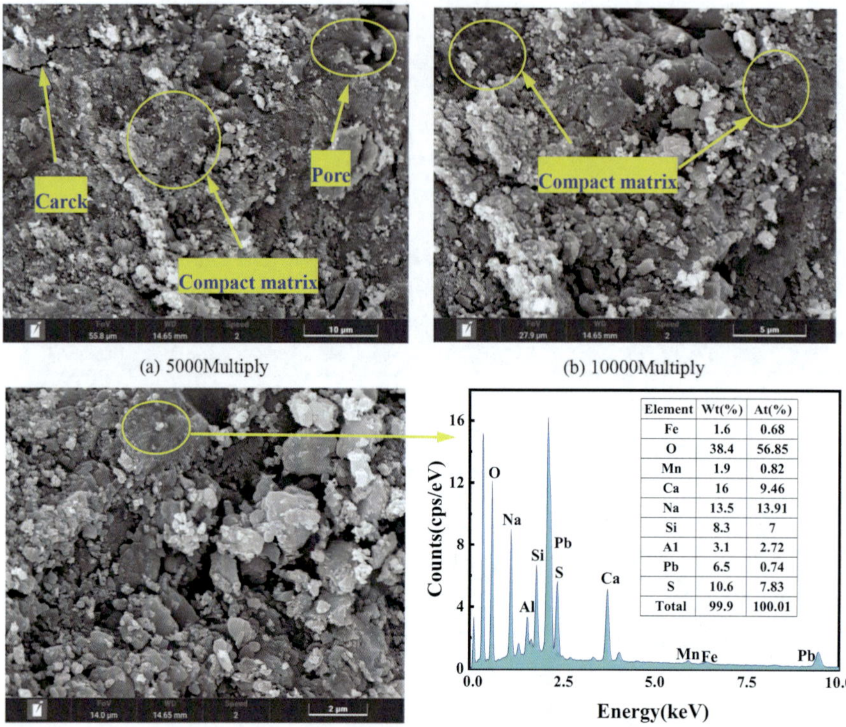

(a) 5000Multiply (b) 10000Multiply

(d) 20000Multiply (d) EDS Result

Fig. 3.24 SEM image and EDS results of EMR (50%) + SS (20%) sample after 28 days. (**a**) 5000 Multiply (**b**) 10,000 Multiply. (**c**) 20,000 Multiply. (**d**) EDS Result

SS specimen, the atomic ratios of Ca/Si, Al/Si, and Na/Si are 1.35, 0.39, and 1.99, respectively. Compared to the 10% EMR + 60% SS specimen, the Ca/Si and Na/Si ratios increase, while the Al/Si ratio decreases. This suggests that when the EMR content reaches 50%, the calcium content in the cementitious system increases, promoting the formation of more C-(A)-S-H and N-A-S-H gels. The Fe content is relatively low, corresponding to the reduced SS content. The sulfur content increases significantly to 7.83%, indicating a further increase in ettringite formation. Similarly, heavy metals are effectively immobilized within the cementitious products.

Figure 3.25 presents the SEM images and EDS results of specimens with different activator concentrations at 28 days. At an activator concentration of 0.1 (Fig. 3.25a), some unhydrated GGBS particles remain in the matrix. This is due to the low alkalinity, which fails to fully activate the reactive components. Consequently, fewer cementitious products are formed in the matrix, insufficient to fill the pores. The connections between cementitious products are weak, resulting in a relatively loose structure with numerous pores, which macroscopically corresponds to lower strength.

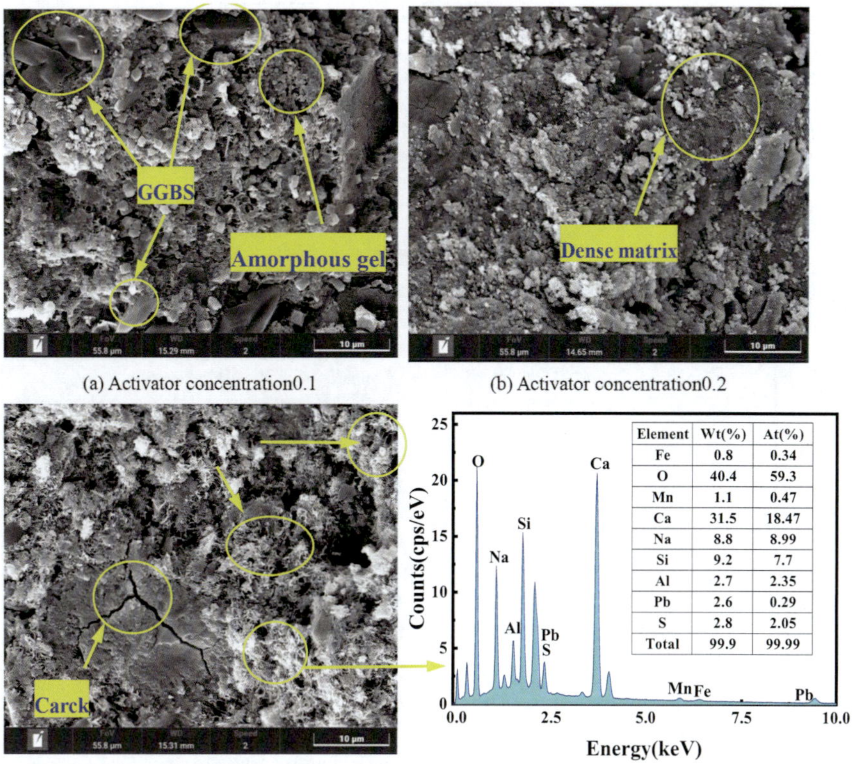

Element	Wt(%)	At(%)
Fe	0.8	0.34
O	40.4	59.3
Mn	1.1	0.47
Ca	31.5	18.47
Na	8.8	8.99
Si	9.2	7.7
Al	2.7	2.35
Pb	2.6	0.29
S	2.8	2.05
Total	99.9	99.99

(a) Activator concentration0.1 (b) Activator concentration0.2 (c) Activator concentration0.3 (d) EDS Result

Fig. 3.25 SEM images (at 5000x magnification) and EDS results of sample after 28 days under different excitation agent concentrations. (**a**) Activator concentration0.1. (**b**) Activator concentration0.2. (**c**) Activator concentration0.3. (**d**) EDS Result

At an activator concentration of 0.2 (Fig. 3.25b), the matrix is the densest, with the most compact connections between cementitious products. Minor cracks are present, but the overall integrity is high, corresponding to the highest macroscopic mechanical strength.

At an activator concentration of 0.3 (Fig. 3.25c), the matrix density is better than that at 0.1, and a significant amount of amorphous network-like gels interlace with each other, contributing to good overall integrity. These amorphous network-like gels fill part of the pores, reducing larger pores and decreasing porosity, which macroscopically results in relatively high mechanical strength. Additionally, larger cracks are observed, likely caused by the more intense hydration reaction at higher activator concentrations.

Figure 3.25d presents the EDS analysis results of the cementitious products. The main elemental composition includes O, Ca, Na, Si, Al, and S. The atomic ratios of Ca/Si, Al/Si, and Na/Si in the gel products are 2.40, 0.31, and 1.17, respectively. The

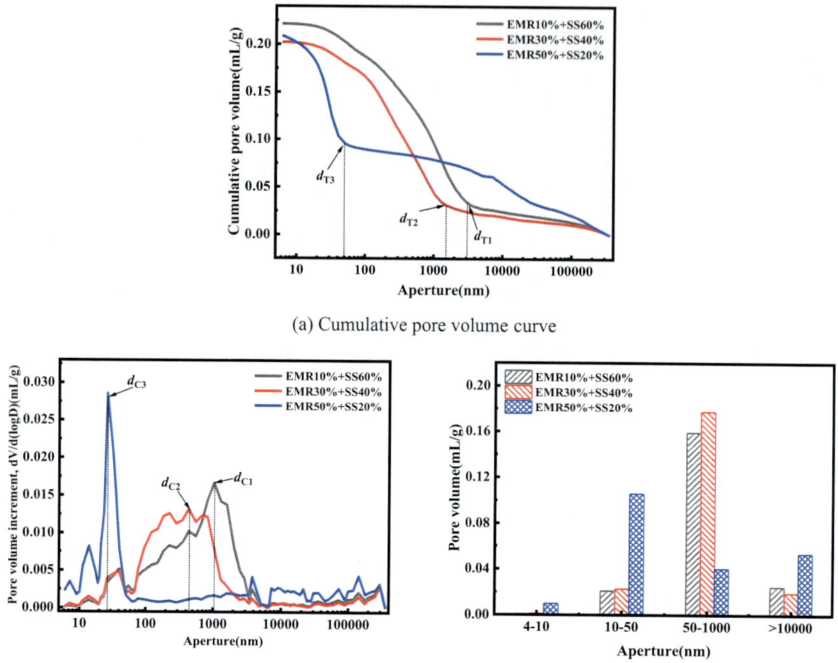

(b) Porosity volume increment distribution curve (c) Statistical analysis of pore volume distribution

Fig. 3.26 Results of mercury porosimetry test on ESGM sample after 28 days. (**a**) Cumulative pore volume curve. (**b**) Porosity volume increment distribution curve. (**c**) Statistical analysis of pore volume distribution

Table 3.3 Pore structure parameters of ESGM sample after 28 days

Sample	Threshold aperture dT(nm)	Critical aperture dC(nm)	Average aperture (nm)	Apparent density (g/mL)	Cumulative pore volume (mL/g)	Porosity (%)
EMR(10%) + SS(60%)	2920.94	1039.67	173.05	2.7474	0.221	37.82
EMR(30%) + SS(40%)	1605.32	434.26	131.40	2.6614	0.202	34.97
EMR(50%) + SS(20%)	55.76	26.29	38.10	2.5407	0.209	34.66

presence of O, Ca, Na, Si, and Al elements indicates the formation of C-(A)-S-H and N-A-S-H gels. The Fe element mainly originates from SS, suggesting that SS participates in hydration to form gels. The high content of S suggests the formation of ettringite within the geopolymer. The presence of Mn and Pb indicates that heavy metals are effectively immobilized in the cementitious products, preventing their leaching.

The MIP test results for specimens cured for 28 days under the conditions of 10% EMR + 60% SS, 30% EMR + 40% SS, and 50% EMR + 20% SS are shown in Fig. 3.26, with the corresponding pore structure parameters listed in Table 3.3. Figure 3.26a displays the cumulative pore volume curves of the ESGM specimens.

It can be observed that the curve for the 10% EMR + 60% SS specimen first shows a sharp increase at a pore diameter of 2920.94 nm. Subsequently, the curves for the 30% EMR + 40% SS and 50% EMR + 20% SS specimens exhibit abrupt increases at pore diameters of 1605.32 nm and 55.76 nm, respectively. These corresponding pore diameter values are defined as the threshold pore diameters (d_T), corresponding to the inflection points on the cumulative pore volume curves in Fig. 3.26a. The results indicate that the d_T is largest for the 10% EMR + 60% SS specimen, followed by the 30% EMR + 40% SS specimen, and smallest for the 50% EMR + 20% SS specimen. This demonstrates that increasing the EMR content significantly reduces the threshold pore diameter of the specimens.

Figure 3.26b shows the incremental pore volume distribution curves of the specimens. The pore diameter corresponding to the peak of each curve is defined as the critical pore diameter (d_C), which represents the largest connected pores within the specimen and is indicated by the maximum incremental pore volume occurring at this diameter. As shown in Fig. 3.26b and Table 3.3, the d_C for the specimen with 10% EMR + 60% SS is 1039.67 nm. This value decreases to 434.26 nm for the 30% EMR + 40% SS specimen, and reaches its minimum of 26.29 nm for the 50% EMR + 20% SS specimen. These characteristic values—the threshold pore diameter (d_T) and the critical pore diameter (d_C)—are closely related to the permeability, diffusivity, and strength of the specimens. Therefore, the MIP results further validate the observed variations in strength. Additionally, Fig. 3.26b indicates that the incremental pore volume distribution curves for the 10% EMR + 60% SS and 30% EMR + 40% SS specimens span a significantly broader range, whereas the curve for the 50% EMR + 20% SS specimen is noticeably narrower, suggesting a lower porosity for the latter.

As listed in Table 3.3, the average pore diameters for the 10% EMR + 60% SS, 30% EMR + 40% SS, and 50% EMR + 20% SS specimens are 173.05 nm, 131.40 nm, and 38.10 nm, respectively. Increasing the EMR content significantly reduces the average pore diameter. The apparent density slightly decreases with higher EMR content, which can be attributed to the higher density of SS compared to that of EMR. The cumulative pore volume is highest for the 10% EMR + 60% SS specimen, while the values for the 30% EMR + 40% SS and 50% EMR + 20% SS specimens are relatively similar. This similarity may be due to the density difference when the SS content is low. For specimens with the same mass, a higher SS content results in a relatively smaller total volume, leading to comparable cumulative pore volumes. The porosities of the 10% EMR + 60% SS, 30% EMR + 40% SS, and 50% EMR + 20% SS specimens are 37.82%, 34.97%, and 34.66%, respectively. As the EMR content increases, the porosity decreases. The mechanical strength of the specimens is strongly correlated with their porosity; generally, a higher porosity corresponds to lower strength. The MIP test results are in good agreement with the strength test results.

According to existing research, the pores within specimens are categorized into four types based on their diameters: small gel pores ($d < 10$ nm), large gel pores (10 nm–50 nm), capillary pores (50 nm–10 μm), and air voids ($d > 10$ μm). Capillary pores form during the early reaction stage of alkali-activated geopolymerization. As hydration progresses, these capillary pores are gradually filled by cementitious products, with some being reduced in size to become gel pores. Gel pores primarily form within gel clusters as gels generated during alkali activation fill larger voids, constituting the main pore structure in alkali-activated cementitious materials.

Figure 3.26c illustrates the pore volume distribution. It can be seen that the pores in the EMR (50%) + SS (20%) specimen are predominantly gel pores, whereas the pores in the EMR (10%) + SS (60%) and EMR (30%) + SS (40%) specimens are mainly capillary pores. The capillary pore volume in the EMR (10%) + SS (60%) specimen is 0.159 mL/g, which decreases to 0.177 mL/g and 0.040 mL/g in the EMR (30%) + SS (40%) and EMR (50%) + SS (20%) specimens, respectively. Conversely, the gel pore volume in the EMR (10%) + SS (60%) specimen is 0.021 mL/g, increasing to 0.023 mL/g and 0.106 mL/g in the EMR (30%) + SS (40%) and EMR (50%) + SS (20%) specimens, respectively. These results indicate that with increasing EMR content, more cementitious products are generated in the specimens. Consequently, a greater number of capillary pores are filled by the gel products and transformed into gel pores, leading to a reduction in capillary pore volume, an increase in gel pore volume, and ultimately a denser microstructure in the specimens.

3.4 Preparation of ECGM: An EMR-Based Grouting Material Incorporating GGBS and Coal Slag

3.4.1 Fresh Properties of the Grout

3.4.1.1 Flowability

Figure 3.27 illustrates the influence of EMR content, activator concentration, and water-to-binder ratio on the flowability of the grout. The results indicate that the flowability increases as the EMR content decreases. When the EMR content is reduced from 70% to 10%, the flowability rises from 161 mm to 251 mm. Notably, at 70% EMR content, the flowability remains below 180 mm. For EMR contents below 25%, the variation in flowability is relatively minor. However, when the EMR content is ≥40%, the change in flowability becomes more significant. At 40% EMR content, the flowability is 211 mm, representing a 31% increase compared to the 70% EMR case. At 10% EMR content, the flowability reaches 251 mm, corresponding to a 55.9% increase relative to the 70% EMR condition. These results demonstrate that EMR content substantially affects grout fluidity: lower EMR content leads to higher fluidity, and vice versa.

The relationship between the water-to-binder ratio and the flowability of ECGM grout is shown in Fig. 3.27b. Considering that the effect of the water-to-binder ratio may vary with different EMR contents, two test groups were set with EMR contents of 55% and 25%, respectively. In both the ECGM 6–11 and ECGM 12–17 groups, the alkali activator content was fixed at 14%, and the GGBS content was 20%. For

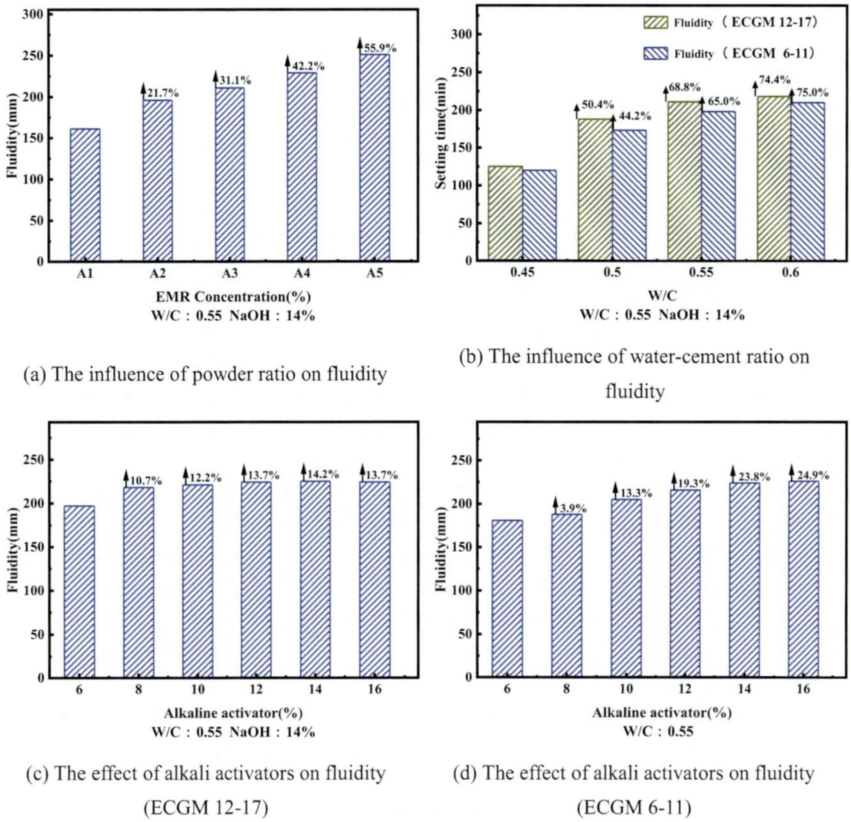

(a) The influence of powder ratio on fluidity

(b) The influence of water-cement ratio on fluidity

(c) The effect of alkali activators on fluidity (ECGM 12-17)

(d) The effect of alkali activators on fluidity (ECGM 6-11)

Fig. 3.27 The influence of material ratio, activator concentration and water-cement ratio on the fluidity of the slurry. (**a**) The influence of powder ratio on fluidity. (**b**) The influence of water-cement ratio on fluidity. (**c**) The effect of alkali activators on fluidity (ECGM 12–17). (**d**) The effect of alkali activators on fluidity (ECGM 6–11)

the ECGM 6–11 group, water-to-binder ratios of 0.45, 0.50, 0.55, and 0.60 correspond to flowability values of 125 mm, 188 mm, 217 mm, and 223 mm, respectively, indicating a positive correlation between the water-to-binder ratio and fluidity. In the ECGM 12–17 group, increasing the water-to-binder ratio from 0.45 to 0.60 resulted in a 74.4% increase in flowability, while the ECGM 6–11 group showed an increase of 88 mm. A higher water-to-binder ratio provides a larger amount of free water for the system, and excessive free water reduces the frictional forces between fine particles, thereby enhancing flowability.

The influence of the alkali activator on the flowability of ECGM grout is presented in Fig. 3.27c, d. To account for potential differences in how activator concentration affects grouts with varying EMR contents, two groups with EMR contents of 55% and 25% were also established. When the NaOH activator content is low, both the ECGM 6–11 and ECGM 12–17 groups exhibit relatively low flowability. As the activator content increases, the flowability of both groups rises, reaching 180 mm. Comparing Fig. 3.27c, d, the increase in flowability with higher activator content is less pronounced for the ECGM 12–17 group, primarily because an excessively high activator content leads to the accumulation of hydration products, which can reduce fluidity. In contrast, the activator concentration has a more significant effect on the flowability of the ECGM 6–11 group. The increase in flowability between activator contents of 6% and 16% is 48 mm, indicating that the activator has a greater impact on fluidity in groups with higher EMR content and a lesser effect in groups with higher CSS (Coal Slag) content. Additionally, GGBS supplies a substantial amount of CaO to the reaction system, promoting the formation of abundant C-(A)-S-H gel during geopolymerization, which also influences flowability [12].

The physical properties of solid powders and the chemical reactions occurring among them contribute to changes in grout fluidity [13, 14]. The finer particles of CSS enhance the fluidity of cementitious materials more effectively than EMR particles. Moreover, differences in particle geometry exist; increasing EMR content reduces flowability. Conversely, reducing EMR content (i.e., increasing CSS content) improves flowability, indicating better fluidity with higher CSS content. A constant water-to-binder ratio of 0.55 was maintained across all test groups. CSS contains a relatively low number of particles reactive in geopolymerization and exhibits weak early-age activity. Therefore, the incorporation of CSS can influence the early-age compressive and flexural strength of the system, affect the reaction rate, and enhance grout fluidity.

Based on the comprehensive analysis of the effects of powder mix proportion, water-to-binder ratio, and alkali activator content on the flowability of the alkali-activated grout, the optimal parameters are determined as follows: an alkali activator content of 14%, a water-to-binder ratio of 0.55, and the powder mix proportion corresponding to ECGM 12–17. Under these conditions, the prepared grout exhibits the best flow performance.

3.4.1.2 Initial and Final Setting Times

The effects of different EMR contents, activator concentrations, and water-to-binder ratios on the setting times of ECGM are shown in Fig. 3.28. Figure 3.28a illustrates the influence of EMR content on the setting times of ECGM. As the EMR content decreases, both the initial and final setting times of the grout are reduced. When the EMR content decreases from 70% to 40% (i.e., from ECGM1 to ECGM3), the initial setting time decreases from 165 min to 121 min (a reduction of 25%), and the final setting time decreases from 182 min to 145 min. This indicates that the grout setting time shortens with a reduction in EMR content. The addition of GGBS also significantly reduces the grout setting time. Incorporating GGBS accelerates the geopolymerization reaction rate, thereby hastening the setting and hardening process of alkali-activated cementitious materials [15]. As indicated in Table 2.1, the high CaO content (34.0%) in GGBS substantially influences the grout setting time.

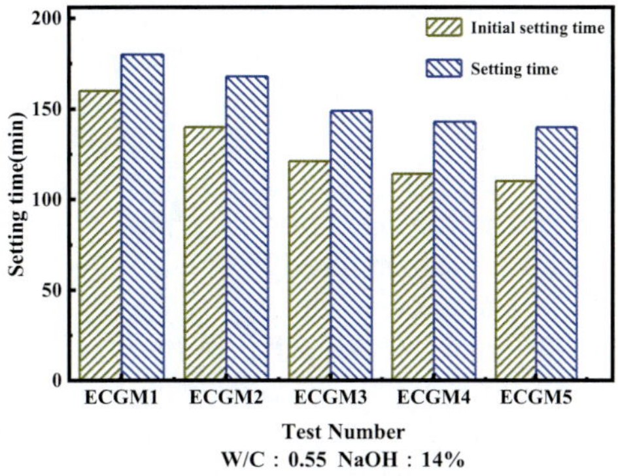

Fig. 3.28 The influence of different powder mix ratios on the setting time. (**a**) Alkali activator (ECGM4). (**b**) Alkali activator (ECGM2)

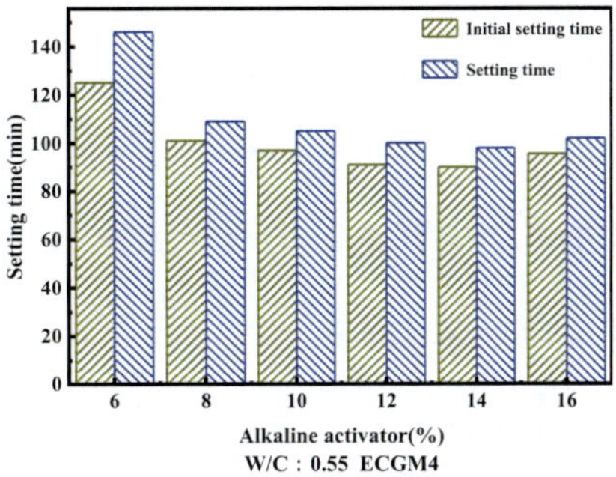

(a) Alkali activator (ECGM4)

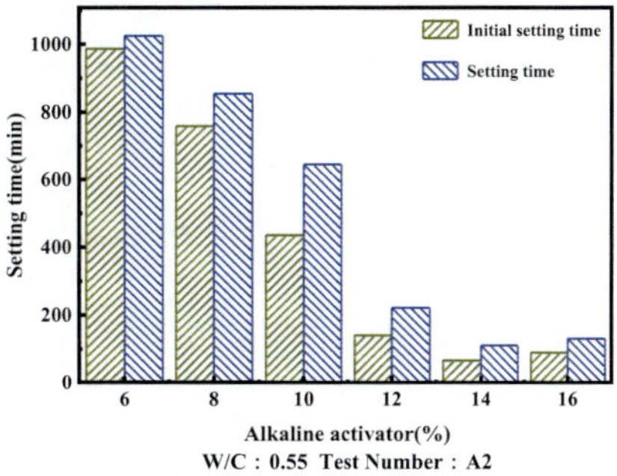

(b) Alkali activator (ECGM2)

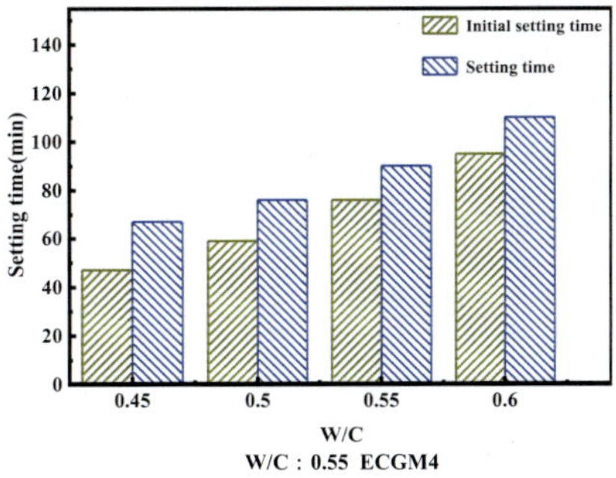

(c) W/C(ECGM4)

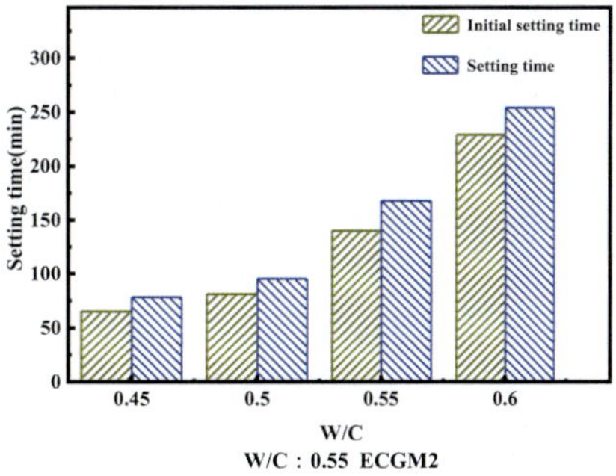

(d) W/C(ECGM2)

Fig. 3.29 The influence of EMR dosage, alkali activator and water-cement ratio on the setting time. (**a**) W/C(ECGM4). (**b**) W/C(ECGM2)

The effects of different alkali activator concentrations on the setting times of ECGM are shown in Fig. 3.29a, b. Figure 3.29a shows that at an alkali activator content of 6%, the initial and final setting times are 125 min and 147 min, respectively. From Fig. 3.29b, it can be observed that when the alkali activator content is below 12%, the setting time of the grout exceeds 600 min. This suggests that a higher EMR content requires a greater amount of alkali activator, and the activator concentration has a more pronounced effect on the setting time of mixtures with higher EMR content. On the other hand, this is because the geopolymerization reaction in the system generates a large amount of gel products. These gels coat the surfaces of unreacted particles, hindering further hydration and consequently prolonging the setting time [7].

The influence of different water-to-binder ratios on the setting times of the ECGM grout is presented in Fig. 3.29c, d. When the water-to-binder ratio increases from 0.40 to 0.60, the initial setting time for the ECGM4 group increases from 47 min to 98 min, while for the ECGM2 group, it increases from 65 min to 229 min. This demonstrates that EMR content significantly affects the grout setting time. The water-to-binder ratio primarily influences the grout's fluidity, which is also largely determined by this ratio. Generally, a higher water-to-binder ratio corresponds to a higher water content in the grout, resulting in relatively longer setting times.

3.4.2 *Strength Characteristics of the Grouting Material*

3.4.2.1 Effect of EMR Content on Compressive Strength

Figure 3.30 shows the effect of different EMR contents on the compressive strength of ECGM. As illustrated in Fig. 3.30a, an appropriate amount of EMR can significantly enhance the mechanical strength of ECGM. When the EMR content is 55%, the 3-day compressive strength reaches 12.4 MPa, while at 25% EMR content, the 3-day compressive strength is 11.1 MPa. The incorporation of EMR causes the

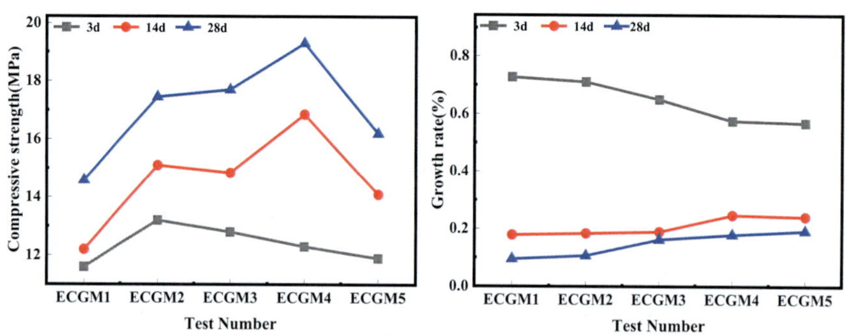

(a) The influence of material mix ratio on compressive strength

(b) Growth rate of compressive strength

Fig. 3.30 The influence of different industrial waste residue powder ratios on compressive strength. (**a**) The influence of material mix ratio on compressive strength. (**b**) Growth rate of compressive strength compressive strength of the hardened specimens to first increase and then decrease. An appropriate dosage of EMR promotes the early-age strength of ECGM, thereby improving its mechanical performance.

Figure 3.30b displays the proportion of the compressive strength increment at each curing age relative to the 28-day compressive strength for different EMR contents. At an EMR content of 55%, the 28-day compressive strength is 15.6 MPa, and the strength increment at 14 days accounts for only 18% of the 28-day strength. In contrast, for the mixture with 25% EMR (ECGM 22–25 group), the 14-day compressive strength reaches 15.9 MPa, with its increment representing 25% of the 28-day strength, which is the highest among all test groups. The 28-day compressive strength of the ECGM 22–25 group is 19.3 MPa, with the strength increment at 28 days accounting for 18% of the total, while for the ECGM 18–21 group, this proportion is only 11%. It is evident that an EMR content of 55% effectively enhances early-age strength, whereas an EMR content of 25% (corresponding to 55% CSS) significantly improves the later-age compressive strength. With increasing CSS content, the hardened specimens exhibit the greatest increase in 28-day compressive strength, indicating that the incorporation of CSS leads to higher strength in the later stages. The reactive components in CSS do not accelerate the hydration reaction in the early stages but rather act as a somewhat "inert" filler, resulting in slower early-age strength development. The high SiO_2/Al_2O_3 ratio in CSS contributes to the lower early-age strength of the hardened specimens, as a high SiO_2/Al_2O_3 ratio in the raw materials of ECGM can negatively affect the compressive strength of cementitious materials [16]. Additionally, the low Al_2O_3 content in the early reaction stage slows down the geopolymerization process, leading to the formation of less C-(A)-S-H gel [14–17]. A higher GGBS content accelerates the geopolymerization rate, promoting the generation of more C-(A)-S-H gel [12–18]. The abundant calcium content in GGBS reacts with other activated particles in the alkaline environment to produce a substantial amount of C-(A)-S-H gel. The formation of C-(A)-S-H gel in the system, in turn, influences the setting and hardening of the cementitious material, further accelerating the geopolymerization rate [19].

3.4.2.2 Effect of Activator Concentration

Figure 3.31a, b illustrate the influence of different alkali activator (NaOH) concentrations on the compressive strength of the ECGM 12–17 and ECGM 6–11 test groups at curing ages of 3, 14, and 28 days. It can be observed that as the activator concentration increases, the compressive strength of the ECGM 12–17 group first increases and then decreases. When the activator concentration is relatively low (<10%), the compressive strength of the ECGM 12–17 group remains comparatively low. This is because a lower activator concentration has a limited effect on the strength of specimens with higher CSS content. Conversely, an increase in activator concentration leads to an initial rise followed by a decline in the 3-day, 14-day, and 28-day compressive strengths of the hardened specimens in the ECGM 6–11 group. At an activator concentration of 14%, the 28-day compressive strength reaches its

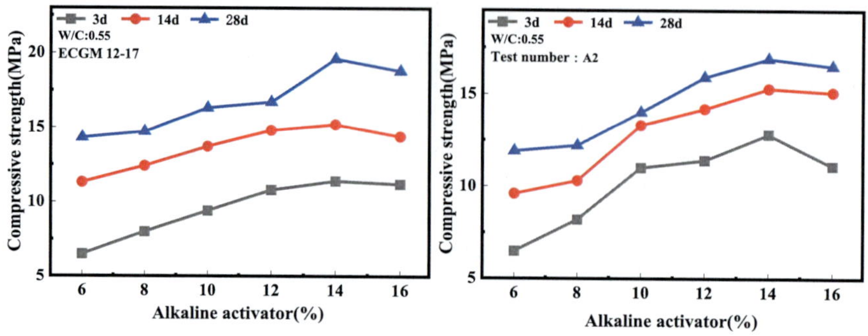

(a) The influence of ECGM 12-17 alkaline activator on compressive strength

(b) ECGM 6-11 - The Effect of Alkali Activators on Compressive Strength

Fig. 3.31 The influence of alkaline activator on compressive strength. (**a**) The influence of ECGM 12–17 alkaline activator on compressive strength. (**b**) ECGM 6–11—The Effect of Alkali Activators on Compressive Strength

maximum value of 16.9 MPa, although it is lower compared to that of the ECGM 12–17 group. The compressive strength of the ECGM 6–11 test group does not vary significantly with changes in activator concentration, indicating that when the EMR content is high, the activator concentration has a minor impact on the strength of the grouting material. On the other hand, when the CSS content is high, the activator concentration plays a more substantial role in determining the strength of the grouting material.

Moreover, when the activator concentration is too low (below 10%), it is insufficient to effectively activate the EMR, CSS, and GGBS. As a result, a portion of the raw materials does not participate in the geopolymerization reaction. These unreacted particles interfere with the formation of hydration products in the system, reducing their yield and consequently lowering the strength of the hardened specimens. Only an appropriate activator concentration can effectively stimulate the raw materials to undergo geopolymerization, producing gel phases with a low calcium-to-silicon ratio, thereby enhancing the mechanical strength of the hardened specimens.

3.4.2.3 Effect of Water-to-Binder Ratio

Figure 3.32a, b show the influence of different water-to-binder ratios on the compressive strength of the ECGM 18–21 and ECGM 22–25 test groups at curing ages of 3, 14, and 28 days. The results indicate an inverse correlation between the water-to-binder ratio and compressive strength. At a water-to-binder ratio of 0.45, the compressive strength of the ECGM 22–25 group reaches 22.8 MPa, and the strength increases significantly with prolonged curing. In contrast, the 28-day compressive

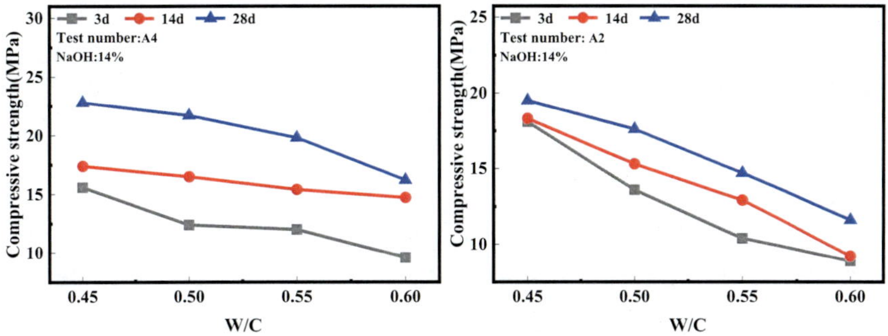

(a) The influence of ECGM 22-25 water-cement ratio on compressive strength

(b) The influence of ECGM 18-21 water-cement ratio on compressive strength

Fig. 3.32 The influence of water-cement ratio on compressive strength. (**a**) The influence of ECGM 22–25 water-cement ratio on compressive strength. (**b**) The influence of ECGM 18–21 water-cement ratio on compressive strength

strength of the ECGM 18–21 group is 19.5 MPa, but the strength increment with curing age is relatively limited.

With a fixed alkali activator content, a lower water-to-binder ratio means a relatively higher proportion of the activator because the amount of water molecules available for the hydration reaction within the system is limited [20]. Conversely, an increased water-to-binder ratio introduces excess water, which hinders the geopolymerization process. Therefore, a higher water-to-binder ratio leads to lower strength. As the water-to-binder ratio increases, the degree of geopolymerization in the grouting material decreases.

3.4.3 *Microstructural Characteristics of the Grouting Material*

3.4.3.1 XRD

Figure 3.33 presents the XRD patterns of ECGM specimens with a curing age of 28 days under different alkali activator concentrations and water-to-binder ratios. As shown in Fig. 3.33a, sharp characteristic diffraction peaks of crystalline phases appear in the XRD patterns. The main crystalline minerals identified are quartz (SiO_2) and calcite. Comparing with the XRD patterns of the raw materials (Fig. 2.6), it can be inferred that minerals such as quartz and anhydrite in the specimens likely originate from CS. As the alkali activator content increases, the peak intensities of crystalline minerals like quartz decrease. The figure also shows that the ECGM 12–17 group exhibits a diffraction pattern similar to that of the ECGM 6–11 test group. However, when the alkali activator concentration exceeds a certain level, the reduction in diffraction peak intensity becomes less pronounced, possibly because

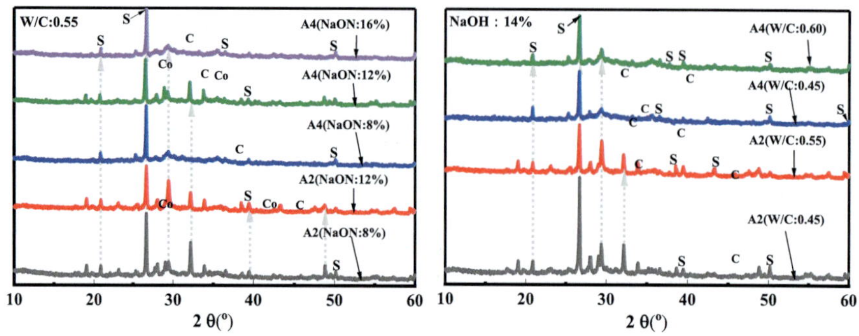

(a) ECGM 6-17(NaOH) (b) ECGM 18-25(W/C)

Fig. 3.33 XRD diffraction spectra of ECGM with different alkali activators and slurry grouting materials with different water-cement ratios(S:SiO_2,C:$Ca_{1.5}SiO_{3.5}xH_2O$,Co:$CaCO_3$). (**a**) ECGM 6–17(NaOH). (**b**) ECGM 18–25(W/C)

the excess activator is not fully consumed in the hydration reaction. A comparison of Fig. 3.33a indicates that, relative to the ECGM 6–11 group, the higher CS content in the ECGM 12–17 group influences the alkali-activated grout system and affects the diffraction peak intensities.

In addition to the crystalline mineral diffraction peaks, the broad amorphous hump in the range of 20–40° (2θ) in the XRD patterns of the ECGM specimens corresponds to the amorphous gel composite products of C-(A)-S-H and N-A-S-H formed during the alkali-activation reaction . Such a hump is also observed in Fig. 3.33, indicating that C-(A)-S-H and N-A-S-H composite gels are generated during the alkali-activated geopolymerization process, contributing to the strength of the specimens. The variation in the area of the amorphous hump is related to the quantity of cementitious products—a larger hump area suggests a greater amount of amorphous gels. Comparing with the XRD patterns of the raw materials, it is evident that the diffraction peaks in the XRD patterns of the hydrated grouting materials are shifted to the right, indicating that geopolymerization occurs within the system under the action of the alkali activator, promoting the reaction of active particles in the raw materials to form a large amount of gel products [21, 22]. Moreover, the characteristics of the gel products are manifested as a hump in the patterns [23, 24]. As seen in Fig. 3.33b, as the water-to-binder ratio decreases, the hump area of the hardened specimens gradually increases, indicating the formation of a substantial amount of gel products. The results demonstrate that when the alkali activator concentration is 14%, the amount of gel products formed in the system increases significantly, and the corresponding strength reaches its maximum value. Conversely, an increase in the water-to-binder ratio of the grout adversely affects the hydration reaction, thereby influencing the mechanical strength of the hardened specimens.

3.4.3.2 FTIR

Figure 3.34a, b present the FTIR spectra of ECGM specimens (Groups 6–25) at a curing age of 28 days under different alkali activator concentrations and water-to-binder ratios. The vibrational absorptions at 945 cm^{-1}, 946 cm^{-1}, and 947 cm^{-1} are attributed to Si–O–T (Si, Al) bonds, indicating the formation of geopolymer gels during the geopolymerization process. The characteristic absorption peaks at 1095 cm^{-1}, 1096 cm^{-1}, 1102 cm^{-1}, and 1103 cm^{-1} correspond to the C–S–H gel bands. A greater amount of C–S–H gel is formed in the matrix of the novel grouting material when the alkali activator content is 14% and the water-to-binder ratio is 0.45 [25].

Additionally, the absorption wavelengths of the Si–O bonds from SiO_2 are observed at 447 cm^{-1} and 420 cm^{-1}. An increase in the alkali activator concentration leads to a gradual increase in wavenumber, with the sample containing 14% alkali activator exhibiting the highest wavenumber. A similar trend is observed at a water-to-binder ratio of 0.45, which can enhance the Si–O tetrahedra within the C–S–H gel structure. As shown in Fig. 3.34b, the absorption band at 1645 cm^{-1} is associated with the stretching vibration of crystalline water, suggesting the presence of hydrocarbon compounds in the reaction products [26]. With a decrease in the water-to-binder ratio, geopolymerization proceeds more thoroughly, and the higher intensity of the absorption band at 945 cm^{-1} in the hardened specimens indicates the generation of more C–S–H gel.

As the alkali activator content increases, the absorption peak of the hardened specimen at 946 cm^{-1} becomes progressively sharper, while the intensities of the diffraction peaks at 1098 cm^{-1} and 945 cm^{-1} gradually weaken. This suggests that more SO_4^{2-} in the system participates in the reaction, leading to the formation of a greater amount of ettringite crystals.

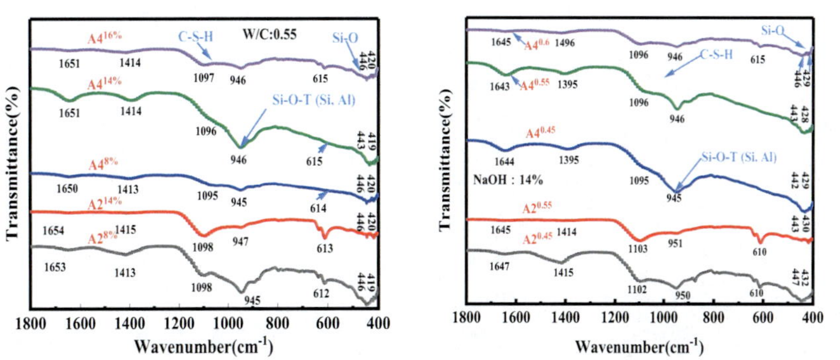

(a)FTIR Analysis of Grouting Materials with Different Alkaline Activators

(b) FTIR Analysis of New Type Grouting Materials with Different Water-Cement Ratios

Fig. 3.34 Chemical structure analysis of samples under different preparation conditions of ECGM 6–25. (**a**) FTIR Analysis of Grouting Materials with Different Alkaline Activators. (**b**) FTIR Analysis of New Type Grouting Materials with Different Water-Cement Ratios

3.4.3.3 SEM-EDS

Microstructural morphology analysis primarily examines the microstructure of specimens from ECGM groups 1–25 after 28 days of curing, aiming to investigate the effects of alkali activator content and water-to-binder ratio on the specimen microstructure.

SEM and SEM-EDS Results of Hardened Specimens with Different Industrial Waste Powder Mix Proportions

Figure 3.35 presents the SEM images and EDS results of hardened specimens from ECGM groups 2 and 4 after 28 days of curing. Cracks, rod-like gels, flaky gels, and flocculent gels can be observed in the images. Under the action of the alkali activator, geopolymerization has generated a substantial amount of gel products with

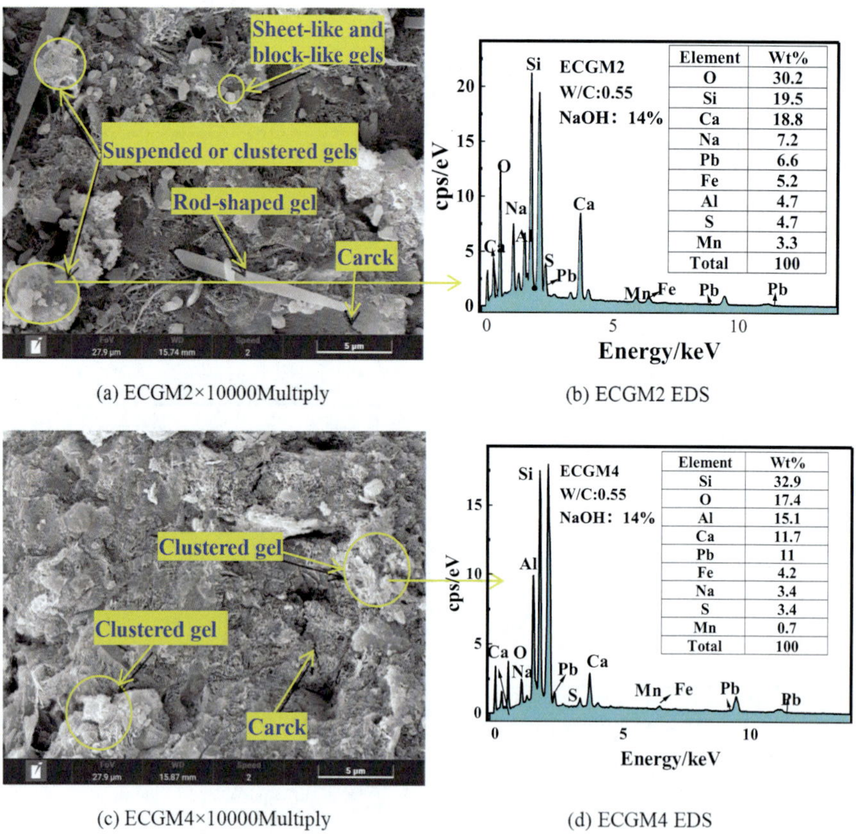

(a) ECGM2×10000Multiply (b) ECGM2 EDS

(c) ECGM4×10000Multiply (d) ECGM4 EDS

Fig. 3.35 SEM images and EDS results of the hardened samples of ECGM2 and ECGM4. (**a**) ECGM2 × 10000 Multiply. (**b**) ECGM2 EDS. (**c**) ECGM4 × 10000 Multiply. (**d**) ECGM4 EDS diverse morphologies within the matrix. Concurrently, it can be seen that the overall structure of the ECGM group 2 specimen is relatively loose and porous, indicating the presence of numerous voids within the matrix that are not completely filled by gel products. In contrast, the ECGM group 4 specimen exhibits a denser and more uniform overall structure compared to group 2. The porous structure within the grouting material can degrade the mechanical performance of the microstructure [27]. The ECGM group 4 specimen shows a denser, more uniform structure with fewer pores relative to the ECGM group 2 specimen.

Figure 3.35b, d present the EDS analysis results of the cementitious products. The main elements detected are Fe, O, Mn, Ca, Na, Si, and Al. The presence of Na indicates that geopolymerization has occurred within the system [28, 29], leading to the formation of typical geopolymeric cementitious material, N-A-S-H gel. GGBS contains a significant amount of Ca, which reacts with Si and Al in an alkaline environment to produce a substantial quantity of C-(A)-S-H gel [29, 30]. Furthermore, under alkaline conditions, GGBS, EMR, and CSS can generate unstable geopolymeric precursors such as $[H_3SiO_4]^-$, $[H_3AlO_4]^{2-}$, and $[Al(OH)_6]^{3-}$ (Eqs. (3.1) to (3.3)) [31]. Influenced by the ambient temperature, these formed precursors undergo further polymerization, resulting in various networked or clustered C-(A)-S-H and N-A-S-H gels (Eqs. (3.4) and (3.5)) [32]. In this process, Ca^{2+} from CSS, EMR, and GGBS first reacts with $[H_3SiO_4]^-$ to form C-S-H gel. Due to differences in chemical bonding, Na^+ reacts with $[H_3SiO_4]^-$ and $[H_3AlO_4]^{2-}$ to generate N-A-S-H gel.

Cementitious products such as C-(A)-S-H gel adhere to the surfaces of low-activity microparticles. The connections within the gel network strengthen over the curing period, and the system continuously rearranges itself, forming a closely interlinked, dense three-dimensional network structure. The formation of mixed gels involves varying degrees of structural order and random dispersion, within which defects such as voids and microcracks can be observed [33]. The C-(A)-S-H gel formed during the geopolymerization process reduces the porosity of the microstructure [34]. Coexisting with N-A-S-H gel, it serves as a crucial source of strength for the specimens. The presence of Mn and Pb indicates that heavy metals are effectively immobilized within the cementitious products, preventing their leaching.

$$SiO_2 + OH^- + H_2O \rightarrow [H_3SiO_4]^- \tag{3.1}$$

$$AlO_2^- + OH^- + H_2O \rightarrow [H_3AlO_4]^{2-} \tag{3.2}$$

$$AlO_2 + OH^- + H_2O \rightarrow [Al(OH)_6]^{3-} \tag{3.3}$$

$$Ca^{2+} + [H_3SiO_4]^- + [H_3AlO_4]^{2-} \rightarrow C-(A)-S-H \tag{3.4}$$

$$Na^+ + [H_3SiO_4]^- + [H_3AlO_4]^{2-} \rightarrow N-A-S-H \tag{3.5}$$

SEM Images and EDS Results of Hardened Specimens Under Different Alkali Activator Concentrations

Figure 3.36 presents the SEM images and EDS results of hardened specimens from the ECGM 6–11 and ECGM 12–17 test groups with alkali activator concentrations of 8% and 16%, respectively. As observed in Fig. 3.36a, c, d, an increase in the alkali activator content leads to the generation of more hydrated gel products within the matrix. However, a comparison with Figure 3.36c reveals a relatively lower amount of gel products in the matrix. When the activator concentration is 8%, the matrix of the ECGM 6–11 test group primarily consists of amorphous network-like gels and contains a substantial number of cracks and pores. Figure 3.36a indicates that under low activator concentration, fewer gel products are formed in the matrix, which is accompanied by numerous cracks and pores. Additionally, some unreacted active particles remain in the specimen under low activator concentration.

Figures 3.36b, f display the EDS detection results of the gel products from ECGM 6–11 and ECGM 12–17, respectively. The results show that the elemental composition of the gel products is consistent with that of the test group with a 14% activator concentration. The EDS results indicate that a lower activator concentration significantly affects the formation of cementitious products, leading to insufficient hydration and consequently reduced strength.

SEM Images and EDS Results of Hardened Specimens with Different Water-to-Binder Ratios

Figure 3.37 presents the SEM images and EDS results of specimens from the ECGM 18–21 and ECGM 22–25 test groups with water-to-binder ratios of 0.45 and 0.6. The results indicate that the matrix is denser at a water-to-binder ratio of 0.45. A lower water-to-binder ratio is more conducive to the formation of gel products. This is particularly evident when comparing the micromorphology in Fig. 3.37a, f, where the lower water-to-binder ratio promotes the generation of a substantial amount of gel products.

A higher water-to-binder ratio results in a greater amount of free water, which further increases the structural porosity and consequently leads to lower compressive strength in the specimens. With a high water-to-binder ratio, the relative number of reactive particles available for geopolymerization is reduced. As a result, the amount of cementitious products formed is insufficient to fill the pores within the system [35, 36]. Additionally, the evaporation of water from the system creates additional voids, further diminishing the compressive strength of the specimens.

In contrast, a lower water-to-binder ratio increases the contact area between reactive particles, enhancing the reaction interface among them. The smaller pore spaces facilitate the formation of abundant gel products, which effectively fill the micropores, thereby contributing to higher specimen strength. Moreover, a lower water-to-binder ratio promotes the generation of more C–S–H gel, which in turn accelerates

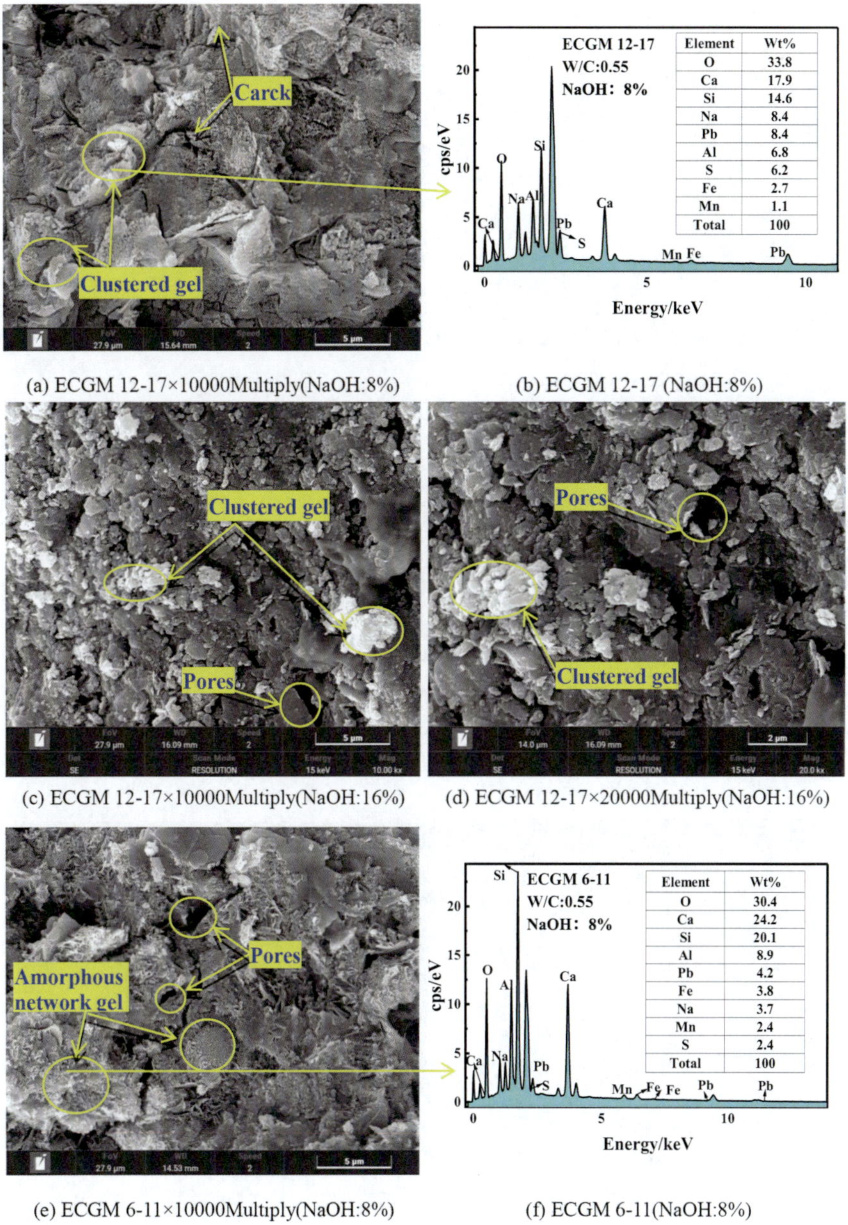

Fig. 3.36 SEM images and EDS results of ECGM 6–11 and ECGM 12–17 hardened specimens (NaOH: 8%, 16%). (**a**) ECGM 12–17 × 10000 Multiply (NaOH:8%). (**b**) ECGM 12–17 (NaOH:8%). (**c**) ECGM 12–17 × 10000 Multiply (NaOH:16%). (**d**) ECGM 12–17 × 20000 Multiply (NaOH:16%). (e) ECGM 6–11 × 10000 Multiply (NaOH:8%). (**f**) ECGM 6–11(NaOH:8%)

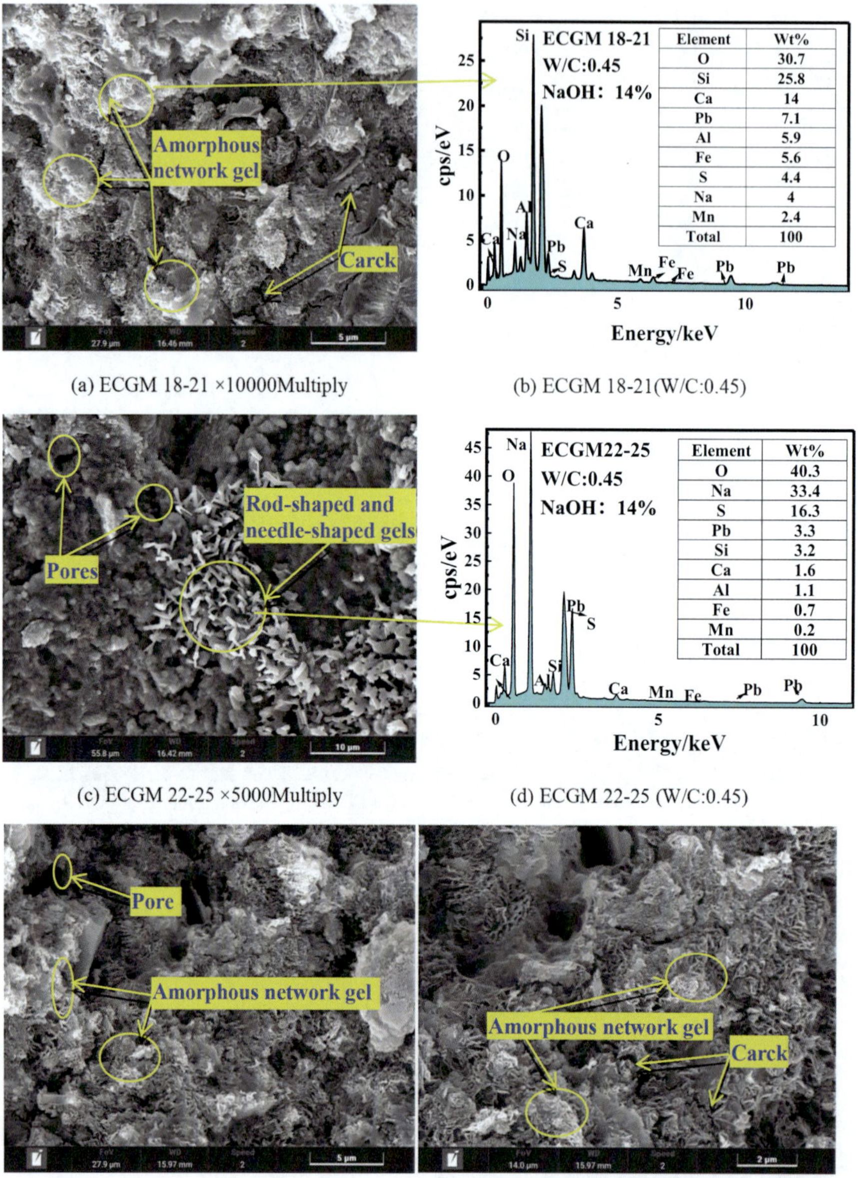

(a) ECGM 18-21 ×10000Multiply

(b) ECGM 18-21(W/C:0.45)

(c) ECGM 22-25 ×5000Multiply

(d) ECGM 22-25 (W/C:0.45)

(e) ECGM 18-21 ×10000Multiply (W/C:0.60)

(f) ECGM 22-25 ×20000Multiply (W/C:0.60)

Fig. 3.37 Scanning electron microscope images and EDS results of hardened specimens ECGM 18–21 and ECGM 22–25 (W/C: 0.45, 0.60). (**a**) ECGM 18–21 × 10000 Multiply. (**b**) ECGM 18–21(W/C:0.45). (**c**) ECGM 22–25 × 5000 Multiply. (**d**) ECGM 22–25 (W/C:0.45). (**e**) ECGM 18–21 × 10000 Multiply (W/C:0.60). (**f**) ECGM 22–25 × 20000 Multiply (W/C:0.60) the dissolution and release of active particles in the system, leading to the formation of a greater amount of N–A–S–H gel [33].

3.4.3.4 MIP

Figure 3.38 presents the MIP test results for specimens from the ECGM 18–21 and ECGM 22–25 test groups after 28 days of curing. Figure 3.38a shows the cumulative pore volume curves of the hardened specimens. It can be observed that for all experimental groups, the cumulative pore volume curve exhibits a sharp decline in the pore diameter range of 45 nm to 1846 nm, followed by a region of gradual decrease (1846 nm–4644 nm). At a pore diameter of 8245 nm, the cumulative pore volume curve for the ECGM 18–21 test group shows a slight increase. Similar phenomena occur at 6542.6 nm for the ECGM 22–25 group (W/C: 0.55) and at 4811.3 nm for the ECGM 22–25 group (W/C: 0.45). The pore diameters at these inflection points are defined as the threshold pore diameters (d_T). Pore diameters beyond this point are negligible in terms of mercury intrusion, meaning that the intrusion volume for pores larger than this diameter is minimal, while the cumulative volume for pores smaller than this point begins to increase sharply.

The results indicate that an increase in CSS content reduces the threshold pore diameter of the specimens. Simultaneously, a lower water-to-binder ratio also leads to a corresponding decrease in the threshold pore diameter. This may be attributed to the irregular, angular shape of CSS particles, which can inhibit shrinkage during internal hydration reactions. Additionally, the gel products formed encapsulate the undissolved CSS microparticles, enhancing the overall compactness and reducing porosity. The ECGM 22–25 group (W/C: 0.45) exhibits the smallest threshold pore diameter, while the ECGM 18–21 group (W/C: 0.55) shows the largest.

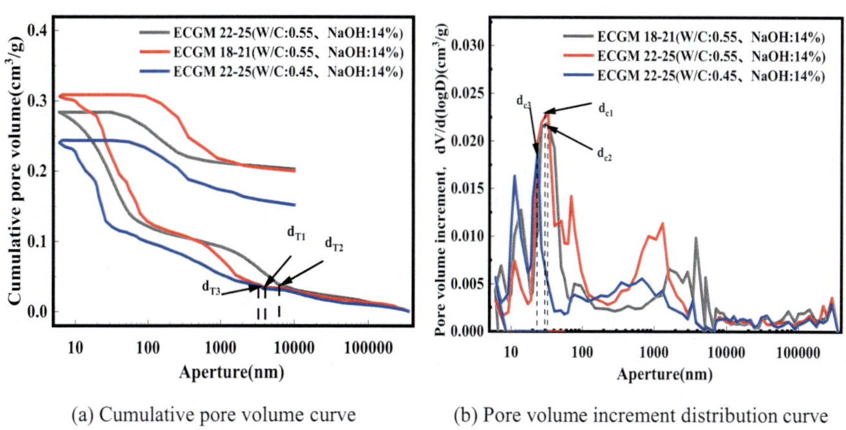

(a) Cumulative pore volume curve (b) Pore volume increment distribution curve

Fig. 3.38 Mercury intrusion test results of hardened specimens of grouting material after 28 days of curing. (**a**) Cumulative pore volume curve. (**b**) Pore volume increment distribution curve

Figure 3.38b shows the incremental pore volume distribution of the specimens. The pore diameter corresponding to the peak in this figure is defined as the critical pore diameter (d_c) [37]. As shown in Fig. 3.38b and Table 3.4, the d_c for the ECGM 20 test group is 66.7 nm, while for the ECGM 24 group (W/C: 0.55) it is 56.7 nm, representing a reduction of 10.3 nm. The critical pore diameter for the ECGM 22 group (W/C: 0.45) further decreases to 43.1 nm, which is 13.6 nm smaller compared to the specimen with a water-to-binder ratio of 0.55. Generally, the values of the critical and threshold pore diameters reflect the permeability of the specimens [37]. Additionally, it can be observed from the figure that the incremental pore volume distribution curve for the ECGM 20 test group spans a significantly wider range, indicating higher porosity in this group. In contrast, the distribution curve for the ECGM 24 group (W/C: 0.55) is narrower than that of the ECGM 20 group, suggesting lower porosity. The curve for the ECGM 22 group (W/C: 0.45) is the narrowest, demonstrating that reducing the water-to-binder ratio of the grout can effectively decrease the porosity of the specimens.

Table 3.4 provides four parameters reflecting the pore structure of the hardened specimens. The data show that the porosity of the ECGM 20 test group is 37.2%, which decreases to 33.3% for the ECGM 24 group (W/C: 0.55). The ECGM 22 group (W/C: 0.45) exhibits an even lower porosity of 30.5%. As indicated in Fig. 3.38a, differences in specimen density can also lead to deviations in the cumulative pore volume. Generally, a higher porosity corresponds to poorer mechanical performance of the specimen [38].

3.5 Chapter Summary

This chapter investigates the activation of the cementitious potential of Electrolytic Manganese Residue (EMR) via high-temperature calcination. The influences of different calcination temperatures on the chemical composition and mineral phases of EMR were characterized using X-ray Fluorescence (XRF), X-ray Diffraction (XRD), and Thermogravimetry-Differential Thermal Analysis (TG-DTA). Strength tests were conducted to evaluate the effect of calcination temperature on the cementitious activity of EMR. Utilizing EMR, Ground Granulated Blast Furnace Slag (GGBS), Fly Ash (FA), Steel Slag (SS), and Coal Slag (CS) as raw materials, with NaOH as the alkali activator, three novel grouting materials—EFGM, ESGM, and ECGM—were developed. The fresh properties (flowability, initial and final setting times) and strength characteristics (uniaxial compressive strength) of these grouts were systematically evaluated. Their microstructural properties were analyzed through XRD, Scanning Electron Microscopy (SEM), Mercury Intrusion Porosimetry (MIP), and Fourier Transform Infrared Spectroscopy (FTIR). The main conclusions are as follows:

1. The primary mineral phases of raw, uncalcined EMR are gypsum ($CaSO_4 \cdot 2H_2O$) and quartz. Calcination at 600 °C leads to the disappearance of the gypsum

Table 3.4 Porosity structure parameters of 28-day hardened specimens of grouting materials

Serial Number	Test materials (g)	NaOH(g)	W/C	Threshold aperture d_T(nm)	Critical aperture d_c(nm)	Average aperture (nm)	Apparent density (g/cm^3)	Porosity (%)
	EMR + CSS + GGBS							
ECGM 20	550 + 250 + 200	140	0.55	8245	66.7	40.16	2.384	37.2
ECGM 24	250 + 550 + 200	140	0.55	6542.6	56.4	26.81	2.365	33.3
ECGM 22	250 + 550 + 200	140	0.45	4811.3	43.1	11.78	2.185	30.5

characteristic peaks and the emergence of anhydrite ($CaSO_4$) peaks. At 800 °C, the intensity of the anhydrite peaks begins to weaken.

2. High-temperature calcination enhances the cementitious activity of EMR. The maximum activity index of 0.8 is achieved at a calcination temperature of 900 °C, identified as the optimal temperature, based on strength development at 3, 14, and 28 days.
3. For all three grouts (EFGM, ESGM, ECGM), an increase in EMR content leads to a decrease in flowability, as does a lower water-to-binder ratio. In ESGM and ECGM systems, a higher alkali activator concentration generally increases flowability. However, in the ECGM system, the activator concentration has a negligible effect on flowability for mixtures with low EMR content, and a decrease is observed at a 16% activator content. The setting times of EFGM first increase and then decrease with higher EMR content. For ESGM and ECGM, setting times decrease with increasing EMR content and higher alkali activator concentration, although excessively high activator concentration in ECGM prolongs the setting times.
4. The strength of all grouting materials decreases with an increasing water-to-binder ratio. EMR content significantly influences strength. In the EFGM system, maximum strength (up to 39 MPa at 28 days) is achieved at 10% EMR content. Strength remains relatively stable (32–34 MPa) within the EMR content range of 20% to 50%, suggesting an EMR content not exceeding 50% is advisable. For ESGM, strength initially increases and then decreases with higher EMR content and activator concentration, peaking at 50% EMR content. Strength development primarily occurs within the 0–14 day curing period. In the ECGM system, strength also shows an initial increase followed by a decrease with increasing EMR content. For mixtures with high EMR content (e.g., ECGM 18–21), compressive strength exhibits minimal variation with changes in activator concentration.
5. The hydration products in the EFGM system include C-(A)-S-H gel, N-A-S-H gel, along with unreacted quartz and mullite, and crystalline phases such as calcium carbonate, ettringite, portlandite, calcium silicate, and sodium sulfate. The primary hydration products in ESGM and ECGM are ettringite, and amorphous C-(A)-S-H and N-A-S-H gels. Increasing EMR content and an appropriate activator concentration promote the formation of more cementitious products, which fill pores and enhance the integrity and density of the microstructure.
6. The water-to-binder ratio significantly affects the pore characteristics. Porosity increases with a higher water-to-binder ratio for all three materials. In the EFGM system, porosity increases with rising EMR content. Conversely, incorporating EMR in the ESGM system effectively reduces the average pore diameter and porosity, generating more cementitious products and resulting in a denser microstructure. In the ECGM system, increasing CSS content effectively lowers the threshold pore diameter, as does a lower water-to-binder ratio. The irregularly shaped CSS particles act as micro-aggregates, mitigating shrinkage during internal hydration. The resulting gel products encapsulate undissolved CSS microparticles, improving overall compactness and reducing porosity.

7. In the ESGM system, C-(A)-S-H and N-A-S-H gels are formed. The intensity of the corresponding absorption peaks increases with higher EMR content, indicating greater gel formation. Increased EMR content also enhances the ettringite absorption peak, signifying more ettringite formation. In the ECGM system, a lower water-to-binder ratio facilitates more complete geopolymerization, increasing C-S-H gel production. With higher alkali activator content, the absorption peak at 946 cm^{-1} in hardened specimens sharpens, indicating that more SO_4^{2-} in the system participates in the reaction to form additional ettringite crystals.

References

1. Wang, J., Peng, B., Chai, L., et al. (2013). Preparation of electrolytic manganese residue-ground granulated blastfurnace slag cement. *Powder Technology, 241*, 12–18.
2. Song, Q., Guo, M., Wang, L., et al. (2021). Use of steel slag as sustainable construction materials: A review of accelerated carbonation treatment. *Resources Conservation and Recycling, 173*, 14.
3. Ling, J. (2022). *Experimental study on performance optimization of mechanical activation coal gangue-based grouting materials*. China University of Mining and Technology.
4. Xiaoyu, Y., Jiawei, W., Haifeng, W., et al. (2022). Study on water absorption rate and saturation coefficient of sintered bricks made from electrolytic manganese slag. *Mineral Resources Comprehensive Utilization, 06*, 89–93.
5. Cui, J., He, Z., & Cai, X. (2022). Mechanical properties and key intrinsic factors of sprayed ultra-high performance concrete (SUHPC) with alkali-free accelerator. *Construction and Building Materials, 349*, 14.
6. Sha, S., Mantellato, S., Weckwerth, S. A., Zhang, Z., Shi, C., Flatt, R. J., et al. (2023). Do superplasticizers work the way we think? New insights from their effect on the percolation threshold of limestone pastes. *Cement and Concrete Research, 172*, 107235.
7. Zhang, W., Li, S., Wei, J., et al. (2018). Grouting rock fractures with cement and sodium silicate grout. *Carbonates and Evaporites, 33*(2), 211–222.
8. Zhang, Y., Yu, P., Pan, F., et al. (2018). The synergistic effect of AFt enhancement and expansion in Portland cement-aluminate cement-FGD gypsum composite cementitious system. *Construction and Building Materials, 190*, 985–994.
9. Chen, M., Wei, J., Zhang, R., et al. (2021). Analysis of basic physical and chemical characteristics of manganese slag before and after solidification and its feasibility as highway slope. *Materials, 14*(19), 5530.
10. Qiang, Z. (2011). *Study on preparation of slag cementitious materials by electrolytic manganese slag-alkali composite excitation*. Central South University.
11. Han, X. (2021). *Alkali-exposed slag and its hydration products on chloride ion solidification performance*. Qingdao University of Technology.
12. Haijun, D. (2014). *Study on properties and microscopic characteristics of composite cement-based-sodium silicate double liquid grouting material*. China University of Geosciences.
13. Mo, L., Lv, L., Deng, M., et al. (2018). Influence of fly ash and metakaolin on the microstructure and compressive strength of magnesium potassium phosphate cement paste. *Cement and Concrete Research, 111*, 116–129.
14. Al-Majidi, M. H., Lampropoulos, A., Cundy, A., et al. (2016). Development of geopolymer mortar under ambient temperature for in situ applications. *Construction and Building Materials, 120*, 198–211.

15. Chindaprasirt, P., Chareerat, T., Hatanaka, S., et al. (2011). High-strength Geopolymer using fine high-calcium Fly ash. *Journal of Materials in Civil Engineering, 23*(3), 264–270.
16. Huseien, G. F., Tahir, M. M., Mirza, J., et al. (2018). Effects of POFA replaced with FA on durability properties of GBFS included alkali activated mortars. *Construction and Building Materials, 175*, 174–186.
17. Puligilla, S., & Mondal, P. (2013). Role of slag in microstructural development and hardening of fly ash-slag geopolymer. *Cement and Concrete Research, 43*, 70–80.
18. Ranjbar, N., Mehrali, M., Alengaram, U. J., et al. (2014). Compressive strength and microstructural analysis of fly ash/palm oil fuel ash based geopolymer mortar under elevated temperatures. *Construction and Building Materials, 65*, 114–121.
19. Ariffin, M. A. M., Hussin, M. W., & Bhutta, M. A. R. (2011). Mix design and compressive strength of Geopolymer concrete containing blended ash from agro-industrial wastes. *Advanced Manufacturing Systems, 339*, 452–457.
20. Li Yanping, S. L. M. R. (2017). A design of experiment approach to study the effects of raw material on the performance of geopolymer concrete. *Advances in Civil Engineering Materials, 6*(1), 20160007.
21. Hojati, M., & Radlinska, A. (2017). Shrinkage and strength development of alkali-activated fly ash-slag binary cements. *Construction and Building Materials, 150*, 808–816.
22. Nan, Y. (2016). *Preparation of Geopolymer and formation strength mechanism of Bayer process red mud activated pretreatment*. Huazhong University of Science and Technology.
23. Ahmari, S., Zhang, L., & Zhang, J. (2012). Effects of activator type/concentration and curing temperature on alkali-activated binder based on copper mine tailings. *Journal of Materials Science, 47*(16), 5933–5945.
24. Andrejkovicova, S., Sudagar, A., Rocha, J., et al. (2016). The effect of natural zeolite on microstructure, mechanical and heavy metals adsorption properties of metakaolin based geopolymers. *Applied Clay Science, 126*, 141–152.
25. Khater, H. M., & Gawaad, H. A. A. E. (2016). Characterization of alkali activated geopolymer mortar doped with MWCNT. *Construction and Building Materials, 102*, 329–337.
26. Ibrahim, M., Johari, M. A. M., Rahman, M. K., et al. (2017). Effect of alkaline activators and binder content on the properties of natural pozzolan-based alkali activated concrete. *Construction and Building Materials, 147*, 648–660.
27. Kujlu, R., Moslemzadeh, M., Rahimi, S., et al. (2020). Selecting the best stabilization/solidification method for the treatment of oil-contaminated soils using simple and applied best-worst multi-criteria decision-making method. *Environmental Pollution, 263*, 11447.
28. Fu, Y., Qiao, H., Feng, Q., et al. (2023). Self-stabilisation of high-temperature calcined electrolytic manganese residue in mortar. *Construction and Building Materials, 386*, 131460.
29. Wang, J., Ma, T., Luan, Y., et al. (2023). Investigation on the effects of desulfurization gypsum on the engineering properties of ternary geopolymers: Improving the utilization of industrial wastes. *Journal of Cleaner Production, 414*, 137638.
30. Gu, F., Xie, J., Vuye, C., et al. (2023). Synthesis of geopolymer using alkaline activation of building-related construction and demolition wastes. *Journal of Cleaner Production, 420*, 138335.
31. Ozcelikci, E., Kul, A., Gunal, M. F., et al. (2023). A comprehensive study on the compressive strength, durability-related parameters and microstructure of geopolymer mortars based on mixed construction and demolition waste. *Journal of Cleaner Production, 396*, 136522.
32. Tan, J., Cai, J., & Li, J. (2022). *Recycling of unseparated construction and demolition waste (UCDW) through geopolymer technology* (Vol. 341, p. 127771). Construction and Building Materials.
33. Shi, Y., Zhao, Q., Xue, C., et al. (2023). Preparation and curing method of red mud-calcium carbide slag synergistically activated fly ash-ground granulated blast furnace slag based eco-friendly geopolymer. *Cement & Concrete Composites, 139*, 104999.

34. He, W., Li, R., Nie, D., et al. (2022). *Belite-calcium sulphoaluminate cement prepared by EMR and BS: Hydration characteristics and microstructure evolution behavior* (Vol. 333, p. 127415). Construction and Building Materials.
35. Tang, L., He, Z., Chen, K., et al. (2023). Study of microscopic properties and heavy metal solidification mechanism of electrolytic manganese residue-based cementitious materials. *Environmental Science and Pollution Research, 30*(48), 105056–105071.
36. Wang, F., Long, G., Bai, M., et al. (2023). A new perspective on Belite-ye'elimite-ferrite cement manufactured from electrolytic manganese residue: Production, properties, and environmental analysis. *Cement and Concrete Research, 163*, 107019.
37. Wang, Q., Han, S., Yang, J., et al. (2023). *Preparation of geopolymer concrete with Bayer red mud and its reaction mechanism* (Vol. 409, p. 133730). Construction and Building Materials.
38. Li, Z., Xiong, Z., Zhang, B., et al. (2023). *Seawater used to Metakaolinite-based geopolymer preparation* (Vol. 392, p. 131816). Construction and Building Materials.

Part III
Regulation of Admixtures on Performance of Grouting Materials

Chapter 4
Effects of Admixtures on the Performance of EMR-Based Grouting Materials

4.1 Introduction

In practical grouting engineering, grouting materials must possess both adequate flowability and sufficient mechanical strength, as these properties are crucial for the smooth implementation of the project and the assurance of final quality. Typically, there exists a trade-off between the mechanical properties and flowability of grouting materials; that is, as flowability increases, mechanical performance tends to decrease [1]. This inverse relationship creates a conflict when trying to meet the performance requirements for engineering applications. Given this situation, to better balance these two performance demands and achieve reasonably satisfactory levels in both for practical engineering, it is often necessary to incorporate a small amount of chemical admixtures into the grouting materials. By leveraging the unique properties and mechanisms of these admixtures, the characteristics of the grout can be modified, thereby mitigating the conflict between the two performance requirements and ensuring the quality and effectiveness of the grouting project [2]·

This study selected two types of admixtures: a polycarboxylate superplasticizer and an alkali-free accelerator. Their dosages were varied within a specified range to investigate the effects of different admixture contents on the performance of the novel grouting material. Strength tests were conducted to examine the influence of admixture content on the mechanical properties of the material. Furthermore, microscopic tests such as XRD and SEM were performed to study the impact of admixture content on the microstructural characteristics of the novel grouting material. The use of the polycarboxylate superplasticizer can improve the flowability, mechanical properties, and durability of the grout. The incorporation of the alkali-free accelerator allows for controlled setting times of the grout and can also enhance its mechanical strength.

Z. Wu et al., *Multi-Source Industrial Solid Waste Synergistically Utilized in Manganese Slag-Based Grouting Materials*,
https://doi.org/10.1007/978-981-92-3021-1_4

4.2 Effect of Admixtures on the Performance of EFGM

4.2.1 Effect of Admixtures on the Fresh Properties of EFGM Grout

4.2.1.1 Flowability

Effect of Polycarboxylate Superplasticizer on the Flowability of EFGM Grout

Figure 4.1 illustrates the effect of different polycarboxylate superplasticizer (PCE) dosages on the flowability of EFGM. As shown in the figure, the flowability of EFGM increases with a higher dosage of PCE. When the EMR content is 10% and the PCE dosage increases from 0.2% to 1.0%, the flowability rises from 164 mm to 208 mm, an increase of 44 mm. Similarly, when the EMR content is 50% and the PCE dosage increases from 0.2% to 1.0%, the flowability increases from 107 mm to 151 mm, also an increase of 44 mm.

The effectiveness of PCE in enhancing grout fluidity stems primarily from its inherent properties as an anionic surfactant. Upon mixing and dissolution into the grout, PCE molecules adsorb onto the surface of the grout material particles [3]. This imparts an electrical charge to the particles, causing them to repel each other due to identical charges. This repulsion releases water trapped between the particles, achieving a water-reducing effect and thereby increasing the flowability of the grout. Notably, the enhancing effect of PCE on flowability is more pronounced at higher EMR contents. This may be attributed to the greater amount of water adsorbed between particles when EMR content is high, which amplifies the effectiveness of the superplasticizer. It is important to note that the addition of a superplasticizer forms an adsorption layer on the particle surfaces, which can influence

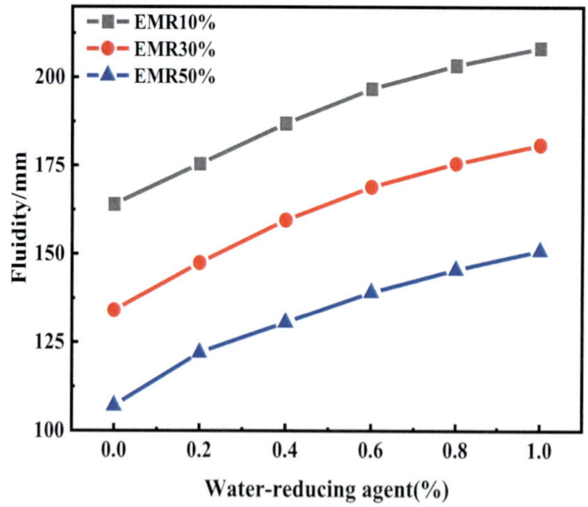

Fig. 4.1 Effect of polycarboxylic acid superplasticizer dosage on EFGM flowability

the hydration process [4]. An excessively high dosage of superplasticizer may lead to over-fluidification of the grout, potentially causing segregation.

Effect of Alkali-Free Accelerator on the Flowability of EFGM Grout

The incorporation of the accelerator significantly affects the flowability. Adding the accelerator effectively reduces the water-to-binder ratio, leading to a noticeable decrease in the flowability of the grout. Moreover, when the water-to-binder ratio is relatively low, the mixing process promotes thorough dissolution of the accelerator, which accelerates the hydration reaction and further impairs flowability [5]. Therefore, during the use of an accelerator, it is advisable to increase the water-to-binder ratio to avoid excessive loss of grout fluidity.

4.2.1.2 Initial and Final Setting Times

Effect of Polycarboxylate Superplasticizer on the Initial and Final Setting Times of EFGM Grout

Figure 4.2 illustrates the influence of different polycarboxylate superplasticizer dosages on the initial and final setting times. As shown, both the initial and final setting times are prolonged with an increasing superplasticizer dosage. Although the setting times indeed extend as the superplasticizer content rises, the extent of this prolongation does not remain constant but rather exhibits a gradually decelerating trend. For instance, when the EMR content is 10%, increasing the superplasticizer

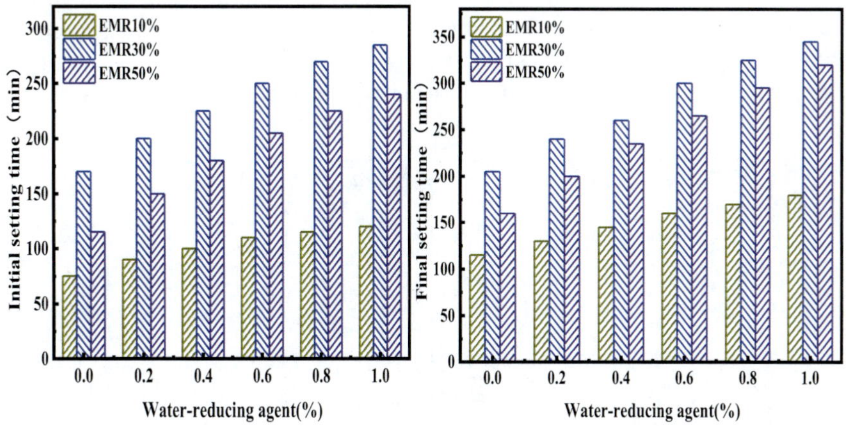

(a) Initial setting time (b) Final setting time

Fig. 4.2 Effect of polycarboxylic acid superplasticizer dosage on initial and final setting times of EFGM. (**a**) Initial setting time. (**b**) Final setting time dosage from 0% to 0.2% extends the initial setting time by 15 min; a further increase to 0.4% prolongs it by an additional 10 min. Similarly, at an EMR content of 30%, raising the superplasticizer dosage from 0% to 0.2% extends the initial setting time by 30 min. When the EMR content is 50%, the same increase in superplasticizer dosage (from 0% to 0.2%) results in a 35-min extension of the initial setting time.

These observations confirm that the retarding effect of the superplasticizer on setting time becomes more pronounced with higher EMR content [6, 7], a trend consistent with its influence on flowability. The superplasticizer adsorbs onto the surface of grout material particles, forming an adsorption film. This film interferes with the normal hydration process, slowing down the reaction rate and consequently extending the setting time [8, 9]. The final setting time follows the same pattern: it is progressively prolonged with higher superplasticizer dosage, and the influencing trend of EMR content aligns with that observed for the initial setting time. Notably, the interval between initial and final setting times widens as the superplasticizer dosage increases, and this effect becomes more evident at higher EMR contents. In summary, the impact of the superplasticizer on setting time is negative; excessive addition may prevent proper setting of the specimen.

Effect of Alkali-Free Accelerator on the Initial and Final Setting Times of EFGM Grout

Figure 4.3 illustrates the effect of the alkali-free accelerator dosage on the initial and final setting times. As the accelerator content increases, both the initial and final setting times of the specimens first decrease and then increase. One key component of the alkali-free accelerator is aluminum sulfate. After being added to the grout, aluminum sulfate rapidly dissolves and ionizes, releasing a large amount of sulfate ions. These ions react with calcium ions in the grout to form secondary gypsum, creating favorable conditions for ettringite formation. This promotes easier generation of ettringite, accelerates the hydration process, and consequently shortens the setting time [8]. However, a higher accelerator dosage is not always better; an

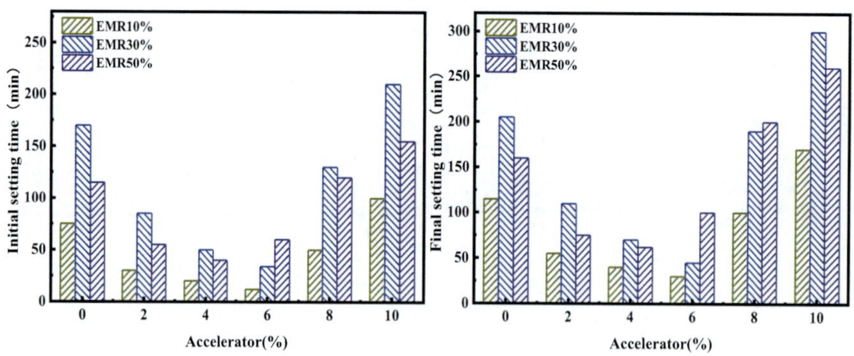

Fig. 4.3 Effect of alkali-free accelerator dosage on initial and final setting times of EFGM

excessive amount can hinder the normal progress of hydration. Overdosing may obstruct the hydration process, leading to prolonged setting times, and in extreme cases, may even prevent the specimen from setting altogether [10, 11].

It can be observed that when the EMR content is 10% and 30%, increasing the accelerator dosage from 0% to 6% results in a shortening trend for both initial and final setting times, with the shortest times achieved at a 6% accelerator content. When the EMR content is 50%, increasing the accelerator dosage from 0% to 4% similarly shows a shortening trend, with the shortest setting times achieved at a 4% accelerator content. This is consistent with the results of the 3-day strength test. This phenomenon occurs because, as the EMR content increases, the amount of sulfate ions released from calcium sulfate ionization also rises. This indicates that the system already possesses relatively sufficient sulfate resources, which can spontaneously accelerate the reaction process, thereby lowering the threshold for the effective accelerator dosage.

4.2.2 *Effect of Admixtures on the Strength Characteristics of EFGM*

4.2.2.1 Effect of Polycarboxylate Superplasticizer on the Strength Characteristics of EFGM

Figure 4.4 illustrates the effect of different polycarboxylate superplasticizer dosages on the compressive strength of EFGM. When the EMR content is 10%, the superplasticizer has a positive effect on the compressive strength of the specimens at the early stage (3 days). However, during the middle stage (14 days), strength development slows significantly; the superplasticizer no longer enhances strength and instead exhibits a negative impact, which becomes more pronounced with increasing dosage. In the later stage, compressive strength increases with the superplasticizer dosage up to 0.6%, reaching its maximum. Beyond this point, further increases in dosage lead to a reduction in strength, with values falling below those of specimens without the superplasticizer.

For specimens with 30% EMR content, the superplasticizer exerts a negative influence on compressive strength during the early and middle stages. In the later stage, as the superplasticizer dosage increases from 0.2% to 0.4%, compressive strength rises, reaching a maximum that exceeds the strength of specimens containing the superplasticizer. However, when the dosage exceeds 0.4%, compressive strength begins to decline.

When the EMR content is 50%, it can be observed that the superplasticizer adversely affects compressive strength throughout the early, middle, and later stages of strength development [12, 13]. The superplasticizer negatively impacts mid-term strength. In the later stage, compressive strength increases with the superplasticizer dosage, peaking at a dosage of 0.8%.

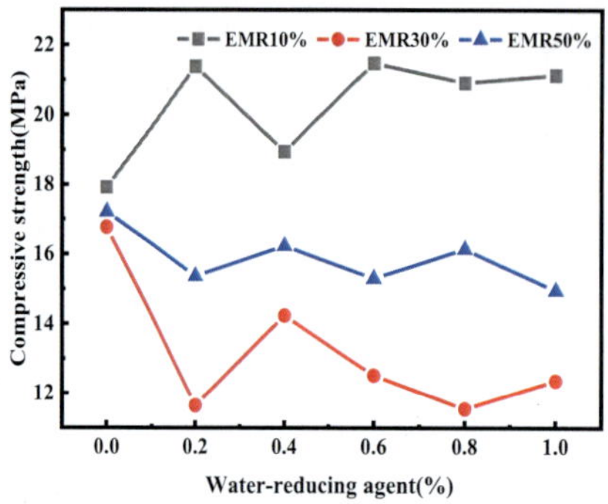

(a) 3-day compressive strength

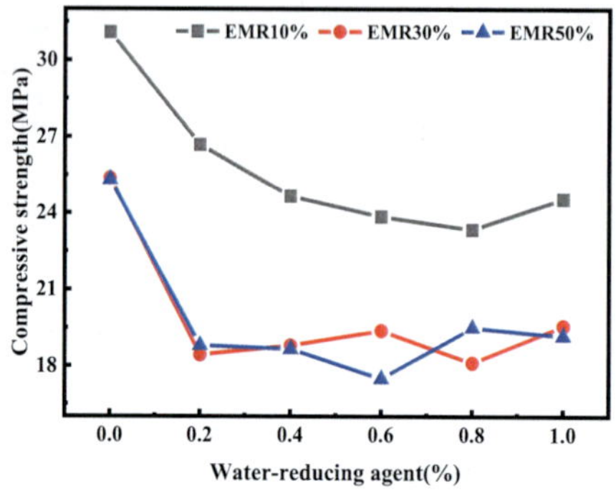

(b) 14-day compressive strength

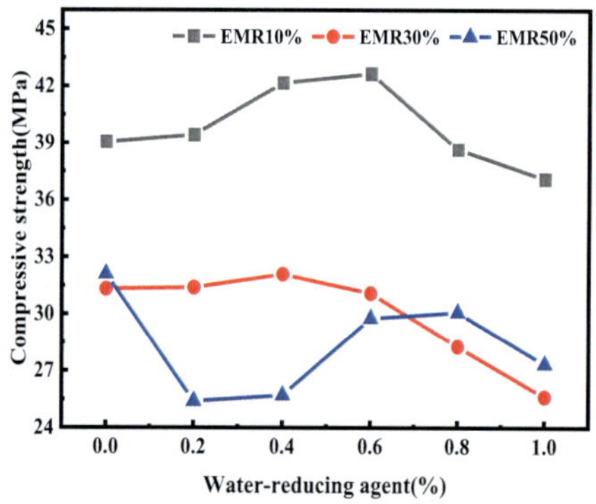

(c) 28d compressive strength

Fig. 4.4 Effect of polycarboxylic acid superplasticizer dosage on compressive strength of EFGM. (**a**) 3-day compressive strength. (**b**) 14-day compressive strength. (**c**) 28d compressive strength

In summary, the optimal superplasticizer dosage is 0.6% for an EMR content of 10%, 0.4% for an EMR content of 30%, and 0.8% for an EMR content of 50%.

4.2.2.2 Effect of Alkali-Free Accelerator on the Strength Characteristics of EFGM

Figure 4.5 illustrates the effect of different dosages of an alkali-free accelerator on the compressive strength. When the EMR content is 10%, the compressive strength increases with higher accelerator dosage, reaching its maximum at a dosage of 4%. Beyond this point, further increases in accelerator content lead to a decline in strength. A particularly negative impact on compressive strength is observed when the accelerator dosage exceeds 6%. For specimens with 30% EMR content, compressive strength similarly increases with accelerator dosage, peaking at 4%.

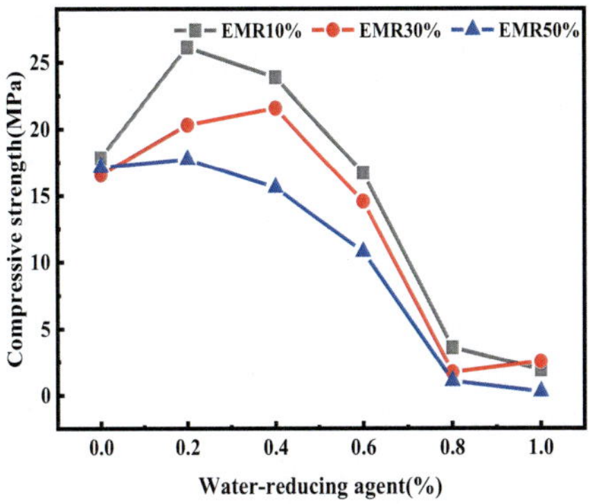

(a) 3-day compressive strength

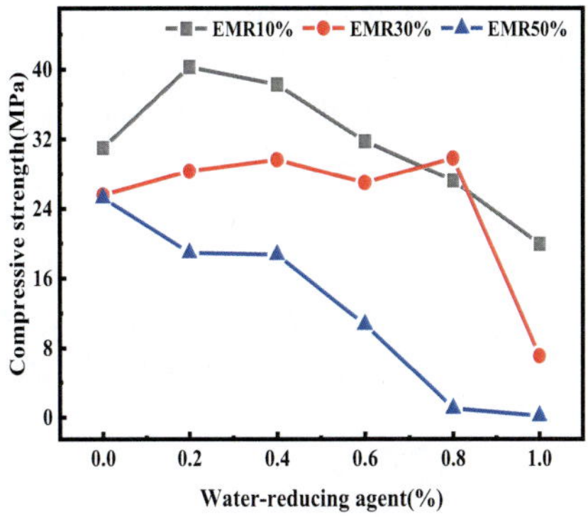

(b) 14-day compressive strength

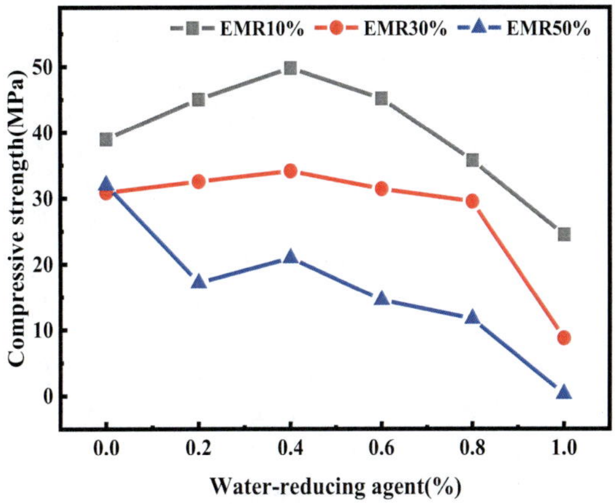

(c) 28d compressive strength

Fig. 4.5 Effect of alkali-free accelerator dosage on EGFM compressive strength. (**a**) 3-day compressive strength. (**b**) 14-day compressive strength. (**c**) 28d compressive strength

Strength decreases with further dosage increases, showing a significant reduction when the dosage surpasses 8%. In the case of 50% EMR content, the accelerator exhibits a clear negative effect on compressive strength. Strength decreases with increasing accelerator dosage, and specimens fail to form properly, showing no measurable compressive strength when the accelerator dosage exceeds 8%.

The alkali-free accelerator contains a substantial amount of soluble aluminum sulfate, which promotes the formation of ettringite during the early hydration stage. This reduces the setting time and enhances early-age strength [14, 15]. When the accelerator dosage does not exceed 4% and the EMR content remains below 30%, a positive effect on later-age strength is also observed. Sulfates additionally activate the GGBS. However, when the accelerator dosage exceeds 6%, the excessive formation of ettringite can hinder later-age strength development. For EMR contents above 30%, the combined effect of the high intrinsic calcium sulfate content from the EMR and the relatively lower GGBS content makes it difficult for the system's active components to counteract and buffer the冲击 of excessive ions. This can prevent proper setting and formation of the EFGM specimens, significantly impairing their later-age strength [16].

4.2.3 *Effect of Admixtures on the Microstructural Characteristics of EFGM*

4.2.3.1 XRD

Figure 4.6 shows the XRD patterns of specimens with a water-to-binder ratio of 0.4, a curing age of 28 days, and an EMR content of 30%, incorporating different dosages of the superplasticizer. The results indicate that the hydration products formed in specimens with varying superplasticizer contents are largely similar. The primary hydration products consist of C-(A)-S-H and N-A-S-H gels, along with unreacted phases such as quartz and mullite, as well as crystalline hydration products including calcium carbonate, ettringite, portlandite, calcium silicate, and sodium sulfate.

The characteristic diffraction humps in the ranges of 5–10° and 20–30° correspond to the C-(A)-S-H and N-A-S-H gels. As the superplasticizer dosage increases, the intensity of these gel-related diffraction features weakens, demonstrating a reduction in the amount of amorphous gels formed. This observation aligns with the trends revealed by the previously conducted strength tests and MIP analysis. When the superplasticizer content reaches 0.8%, characteristic peaks of ettringite appear. This suggests that an excessive amount of superplasticizer hinders the hydration process, preventing ettringite from participating in secondary hydration reactions and thereby affecting the performance optimization of the material system [17–19]. At this superplasticizer dosage, characteristic peaks of portlandite are also observed [20], which is consistent with the findings from SEM analysis.

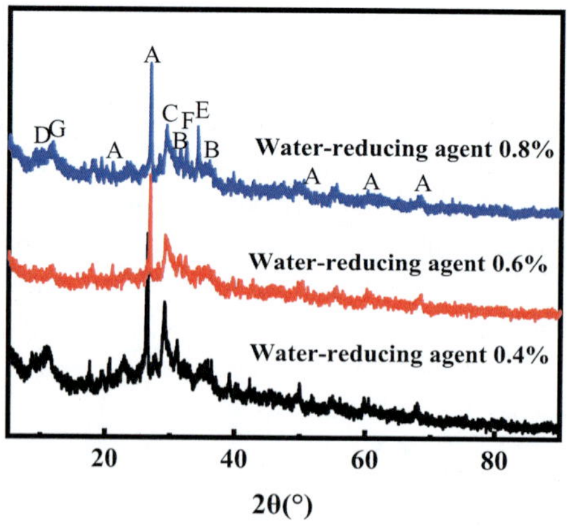

Fig. 4.6 XRD spectrum analysis of samples with different water-reducing agent content (A: Quartz-79-1906, B: Mullite-79-1453, C: Calcium Carbonate-72-1937, D: Calcium Aluminate-41-1451, E: Calcium Hydroxide-87-0673, F: Calcium Silicate-23-1042, G: Sodium Sulfate-99-0103)

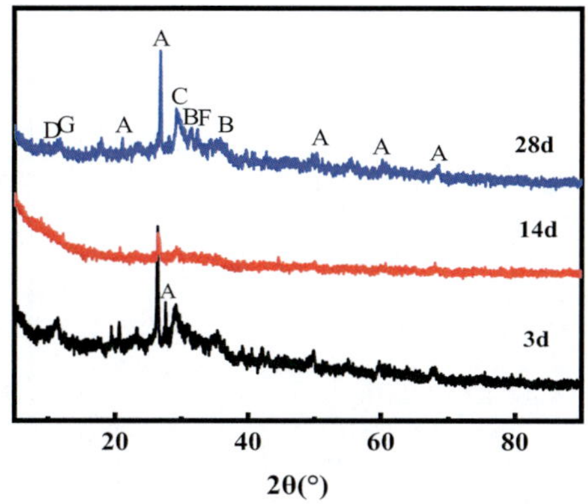

Fig. 4.7 XRD spectrum analysis of samples at different maturity stages (A: Quartz-79-1906, B: Mullite-79-1453, C: Calcium Carbonate-72-1937, D: Calcium Aluminate-41-1451, F: Calcium Silicate-23-1042, G: Sodium Sulfate-99-0103)

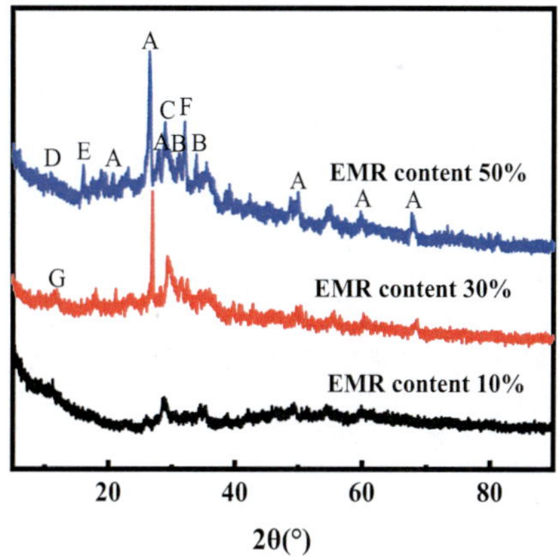

Fig. 4.8 XRD pattern analysis of samples with different manganese slag contents (A: Quartz-79-1906, B: Mullite-79-1453, C: Calcium Carbonate-72-1937, D: Calcium Aluminate-41-1451, E: Calcium Hydroxide-87-0673, F: Calcium Silicate-23-1042, G: Sodium Sulfate-99-0103)

Figure 4.7 presents the XRD patterns of specimens with a water-to-binder ratio of 0.4, an EMR content of 30%, and a superplasticizer dosage of 0.6% at different curing ages. As the curing age increases, the diffraction features corresponding to C-(A)-S-H and N-A-S-H gels in the ranges of 5–10° and 20–30° become more pronounced, indicating a continuous increase in the amount of hydration products. At the later hydration stage (28 days), characteristic peaks of hydration products such as sodium sulfate and calcium silicate appear. This is attributed to the combination of unreacted sulfate ions and sodium ions in the system, leading to the formation of sodium sulfate crystals [21, 22].

Figure 4.8 shows the XRD patterns of specimens with a water-to-binder ratio of 0.4, a curing age of 28 days, and a superplasticizer dosage of 0.6%, under different EMR contents. When the EMR content is 10%, the hydration products mainly consist of amorphous gels such as C-(A)-S-H and N-A-S-H gels. At an EMR content of 30%, characteristic peaks of sodium sulfate appear. When the EMR content reaches 50%, characteristic peaks of ettringite, portlandite, mullite, and other phases are observed. This is because at a superplasticizer dosage of 0.6%, the superplasticizer exerts a significant hindering effect on the hydration process, which also explains

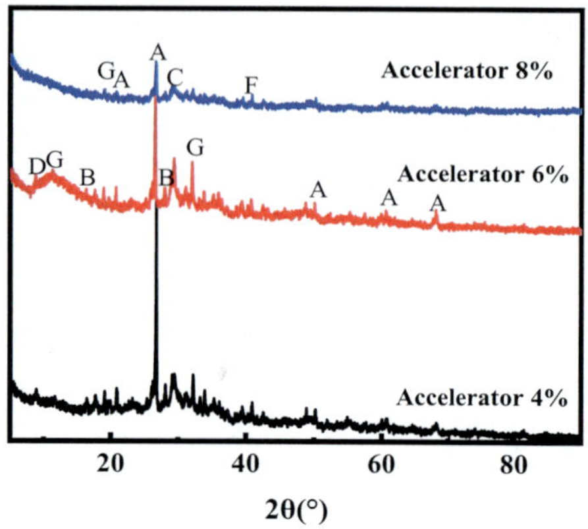

Fig. 4.9 XRD pattern analysis of samples with different manganese slag contents (A: Quartz-79-1906, B: Mullite-79-1453, C: Calcium Carbonate-72-1937, D: Calcium Aluminate-41-1451, E: Calcium Hydroxide-87-0673, F: Calcium Silicate-23-1042, G: Sodium Sulfate-99-0103)

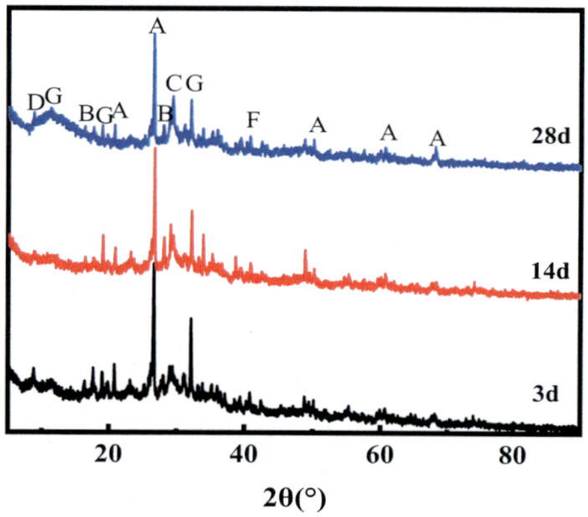

Fig. 4.10 XRD spectrum analysis of specimens at different curing ages (A: Quartz-79-1906, B: Mullite-79-1453, C: Calcium Carbonate-72-1937, D: Calcium Aluminate-41-1451, E: Calcium Hydroxide-87-0673, F: Calcium Silicate-23-1042, G: Sodium Sulfate-99-0103)

why the strength continuously decreases with increasing EMR content under this dosage condition.

Figure 4.9 presents the XRD patterns of specimens with a water-to-binder ratio of 0.4, an EMR content of 30%, and a curing age of 28 days, under different dosages of the accelerator. The diffraction features in the ranges of 5–10° and 20–30° correspond to C-(A)-S-H and N-A-S-H gels. As the accelerator dosage increases, the intensity of these gel-related diffraction features weakens, indicating a reduction in the amount of geopolymer gels formed during the polymerization reaction. This suggests that an excessive amount of accelerator interferes with and hinders the later-stage hydration process, which is consistent with the strength test results [23, 24]. When the accelerator dosage reaches 6%, the intensities of the characteristic peaks for ettringite and sodium sulfate show an increase. This indicates that, at this accelerator dosage, the secondary hydration reaction encounters significant hindrance.

The water-cement ratio was 0.4, with an EMR content of 30% and a rapid setting agent content of 6%. The XRD patterns of specimens at different curing ages are shown in Fig. 4.10. As the curing age increased, the characteristic peaks of sodium

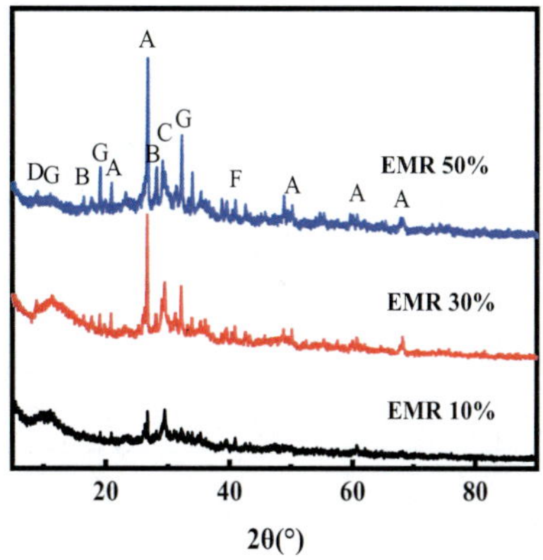

Fig. 4.11 XRD pattern analysis of samples with different electrolytic manganese slag contents (A: Quartz-79-1906, B: Mullite-79-1453, C: Calcium Carbonate-72-1937, D: Calcium Aluminate-41-1451, E: Calcium Hydroxide-87-0673, F: Calcium Silicate-23-1042, G: Sodium Sulfate-99-0103)

sulfate gradually weakened, indicating the continuous consumption of sulfate ions. The figure indicates that the hydration products formed at different curing ages show little variation. The primary hydration products are C-(A)-S-H and N-A-S-H gels, along with unreacted quartz, mullite, and hydration products such as calcium carbonate, calcium aluminate hydrate, calcium silicate, and sodium sulfate.

Figure 4.11 presents the XRD patterns of specimens with a water-to-binder ratio of 0.4, a curing age of 28 days, and an accelerator dosage of 6%, under different EMR contents. As the EMR content increases, the diffraction features of C-(A)-S-H and N-A-S-H gels in the ranges of 5–10° and 20–30° progressively weaken, indicating a reduction in the amount of geopolymer gels formed. This observation aligns with the strength test results, which show a continuous decrease in strength with higher EMR content.

When the EMR content is 10%, the main characteristic peaks observed are those of quartz and calcium carbonate. At an EMR content of 30%, characteristic peaks of ettringite, sodium sulfate, and mullite begin to appear. When the EMR content reaches 50%, the characteristic peaks of sodium sulfate, mullite, quartz, and other phases intensify, indicating an increase in these phases. The abundant presence of these phases can lead to a more complex and unstable internal structure, contributing to the continuous decline in material strength [25, 26].

4.2.3.2 FTIR

Figure 4.12 presents the FTIR spectra of the manganese residue-slag-fly ash-based grouting material with the addition of admixtures. As the accelerator dosage increases from 4% to 8%, the broad O–H stretching vibration band near 3458 cm^{-1} shows no significant shift, with only minor changes in peak intensity, primarily originating from hydrates within the system. The H–O–H bending vibration near 1650 cm^{-1} reflects information about the vibrations within the aluminosilicate network. A higher accelerator dosage promotes the alkali-activated geopolymerization, and the enhanced peak intensity confirms that the accelerator accelerates the

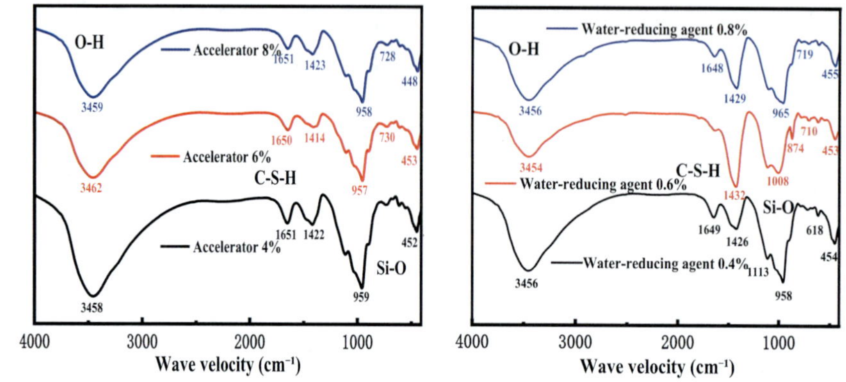

(a) Effect of Accelerators on Grouting Materials (b) Effect of Water Reducers on Grouting Materials

Fig. 4.12 Infrared spectra of grouting materials based on electrolytic manganese slag-slag-fly ash with added admixtures. (**a**) Effect of accelerators on grouting materials. (**b**) Effect of water reducers on grouting materials

precipitation of cementitious phases. The asymmetric stretching vibration peak in the range of 1414–1423 cm^{-1} corresponds to carbonate groups, stemming from carbonate impurities in the materials and carbonation reaction products. An increase in accelerator dosage reduces the sharpness of this peak, indicating a decrease in its crystallinity. The characteristic peak near 728 cm^{-1} is associated with the characteristic vibration of quartz. Its shift toward higher wavenumbers suggests that an increase in accelerator dosage enhances the degree of polymerization of Si–O bonds, leading to increased crystallinity of the hydration products.

Figure 4.12b illustrates the influence of the superplasticizer on the FTIR spectra of the grouting material. It can be observed that at 3456 cm^{-1}, 1426 cm^{-1}, and 958 cm^{-1}, the peaks for specimens with lower superplasticizer dosage are sharper compared to those with higher dosage. At 1426 cm^{-1}, the peak intensity is highest at a superplasticizer content of 0.6 wt.%, indicating that an appropriate dosage of superplasticizer can improve the fluidity of the grout, promote more complete hydration within the system, and reduce the residues of free ions. In contrast, an excessive dosage can lead to competitive adsorption of molecules, inhibiting the active sites of hydration reactions and thereby reducing the amount of hydration products. The intensity of the Si–O characteristic peak near 958 cm^{-1} decreases with increasing superplasticizer dosage, suggesting that more silicate species participate in the geopolymerization reaction [27, 28].

4.2.3.3 SEM-EDS

Figure 4.13 shows the micromorphology of specimens with a water-to-binder ratio of 0.4, an EMR content of 30%, and a curing age of 28 days, under different superplasticizer dosages. As can be seen from the figure, some unreacted spherical FA

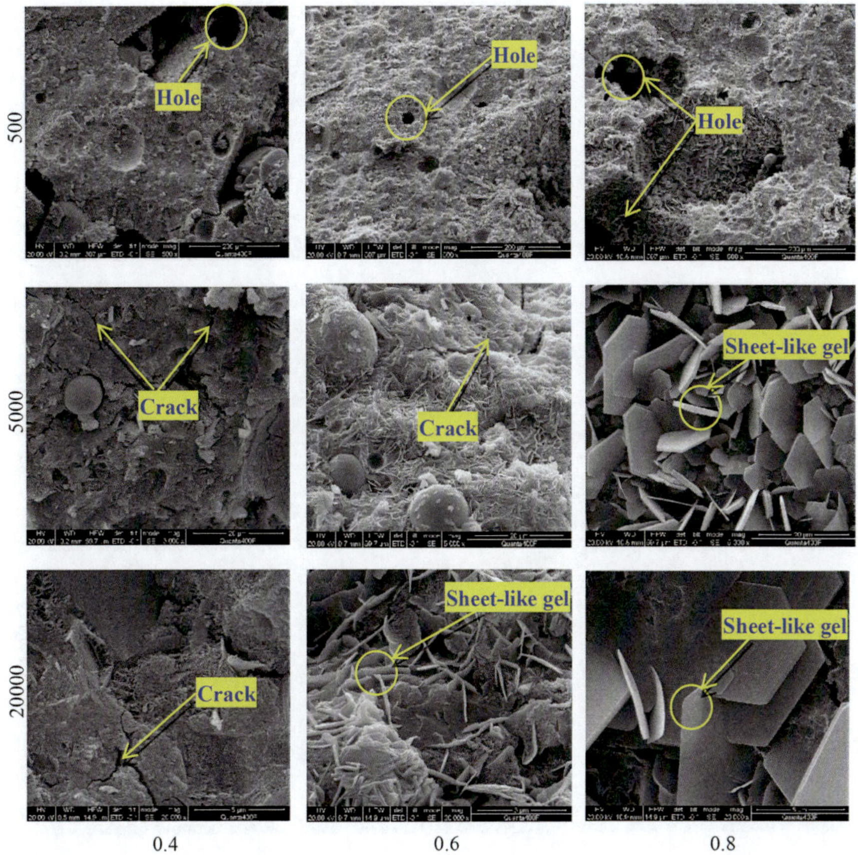

Fig. 4.13 Microstructural morphology of specimens with different water-reducing agent content

particles remain in the hydration products. Under a low magnification of 500×, specimens with different superplasticizer dosages appear very similar in terms of compactness, showing almost no significant differences. However, the situation is quite different under a high magnification of 20,000×. With an increase in superplasticizer dosage, the specimens gradually become looser. Key parameters such as porosity and pore size distribution obtained from MIP tests also confirm that as the superplasticizer dosage increases, the compactness of the specimens decreases and the structure tends to become more porous.

At magnifications of 5000× and 20,000×, when the superplasticizer dosage is 0.6%, a large number of flaky sodium sulfate crystals can be observed. When the dosage increases to 0.8%, numerous hexagonal portlandite crystals are visible. These observations are highly consistent with the XRD results, indicating that when the superplasticizer dosage exceeds 0.4%, the formation of geopolymer gels is hindered. Excessive superplasticizer has a negative effect on both the

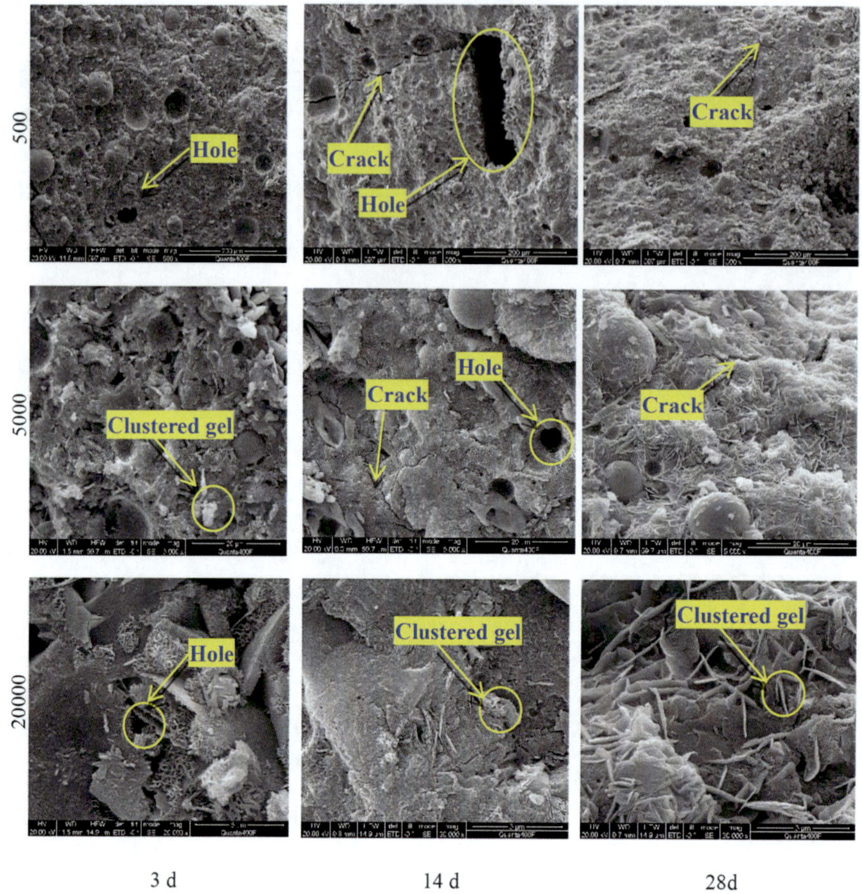

Fig. 4.14 Microstructural morphology of specimens at different curing ages

geopolymerization reaction and the material strength, which aligns with the trends observed in the strength tests [29–31].

Figure 4.14 presents the micromorphology of specimens with a water-to-binder ratio of 0.4, an EMR content of 30%, and a superplasticizer dosage of 0.6%, at different curing ages. Unreacted fly ash (FA) particles can be observed under microscopic examination. As the curing age increases, the amount of hydration products formed increases, and the microstructure of the specimens becomes denser.

Figure 4.15 presents the micromorphology of specimens with a water-to-binder ratio of 0.4, an EMR content of 30%, and a superplasticizer dosage of 0.6%, at different curing ages. Unreacted fly ash (FA) particles can be observed under microscopic examination. As the curing age increases, the amount of hydration products formed increases, and the microstructure of the specimens becomes denser.

Figure 4.16 shows the micromorphology of specimens with a water-to-binder ratio of 0.4, an EMR content of 30%, and a curing age of 28 days, under different

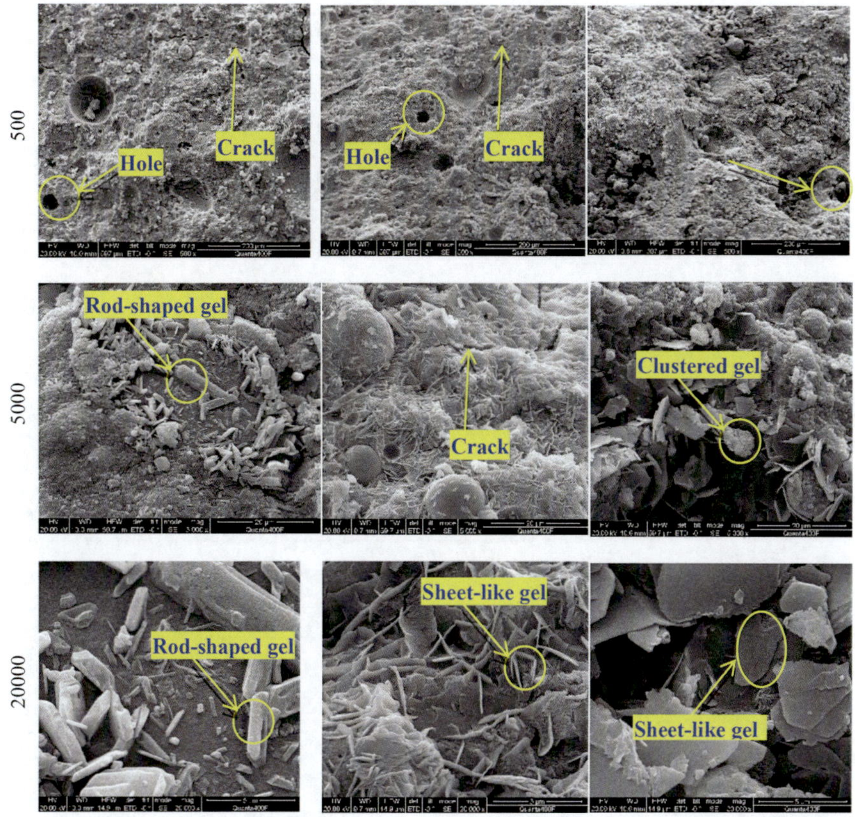

Fig. 4.15 Microstructural morphology of samples with different electrolytic manganese slag content

dosages of the accelerator. As the accelerator dosage increases, the compactness of the microstructure of the specimens shows little significant change, which is consistent with the results from MIP analysis. At an accelerator content of 4%, microcracks and a network-like hydration product can be observed under a magnification of 20,000×. When the accelerator dosage is 6%, pore structures and stacked sodium sulfate crystals are visible under 20,000× magnification. At an accelerator dosage of 8%, flocculent gels are observed, aligning with the XRD results.

Figure 4.17 shows the micromorphology of specimens with a water-to-binder ratio of 0.4, an EMR content of 30%, and an accelerator dosage of 6%, at different curing ages. Although the curing age varied, the micromorphology of the hydration products remained largely consistent across different ages. All specimens exhibited micro-pores, microcracks, stacked flaky sodium sulfate crystals, rod-like quartz, and flocculent geopolymer gels.

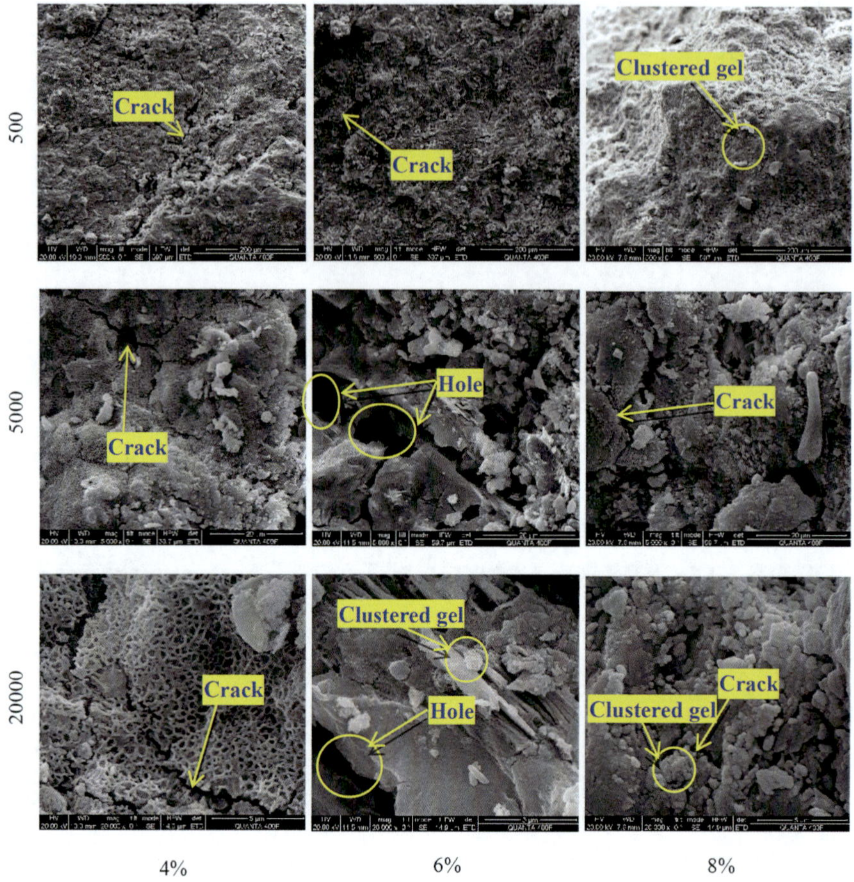

Fig. 4.16 Microstructural morphology of specimens with different accelerator content

Figure 4.18 presents the micromorphology of specimens with a water-to-binder ratio of 0.4, a curing age of 28 days, and an accelerator dosage of 6%, under different EMR contents. When the EMR content is 10%, distinct microcracks are observed. These cracks likely resulted from volumetric changes caused by the reaction of materials encapsulated by hydration products formed under the influence of the accelerator. At an EMR content of 30%, stacked flaky sodium sulfate crystals and pore structures are visible under a magnification of 20,000×. When the EMR content is 50%, flocculent gels and flaky hydration products can be observed. It is evident that as the EMR content increases, the microstructure of the specimens becomes progressively looser. This observation is corroborated by the MIP test results, which show a continuous increase in specimen porosity with higher EMR content.

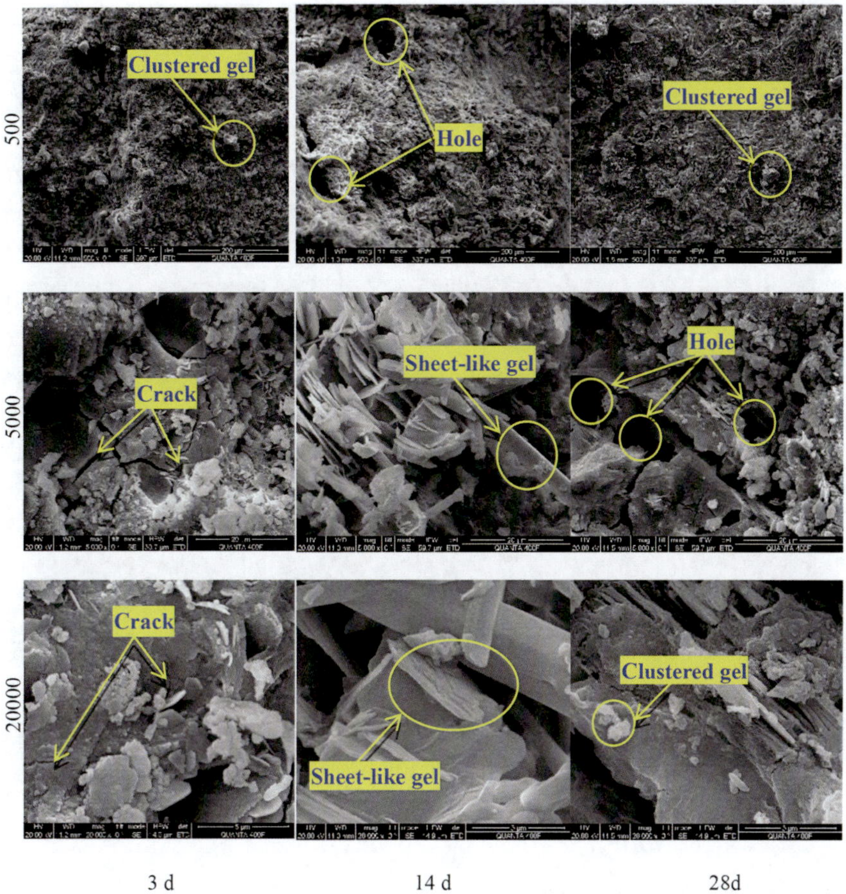

Fig. 4.17 Microstructural morphology of specimens at different ages

4.2.3.4 MIP

Polycarboxylate Superplasticizer (PCE)

The relevant parameters of specimens with different superplasticizer dosages are listed in Table 4.1. As the superplasticizer dosage increases, both the porosity and average pore diameter of the specimens increase; the apparent density first decreases and then increases; and the permeability first increases and then decreases.

The influence of different superplasticizer dosages on the pore structure of the specimens is shown in Fig. 4.19. With an increase in superplasticizer dosage, the cumulative pore volume increases, causing the corresponding curve to shift upward. However, the incremental pore volume remains largely similar in terms of its distribution pattern, though the overall pore volume becomes greater.

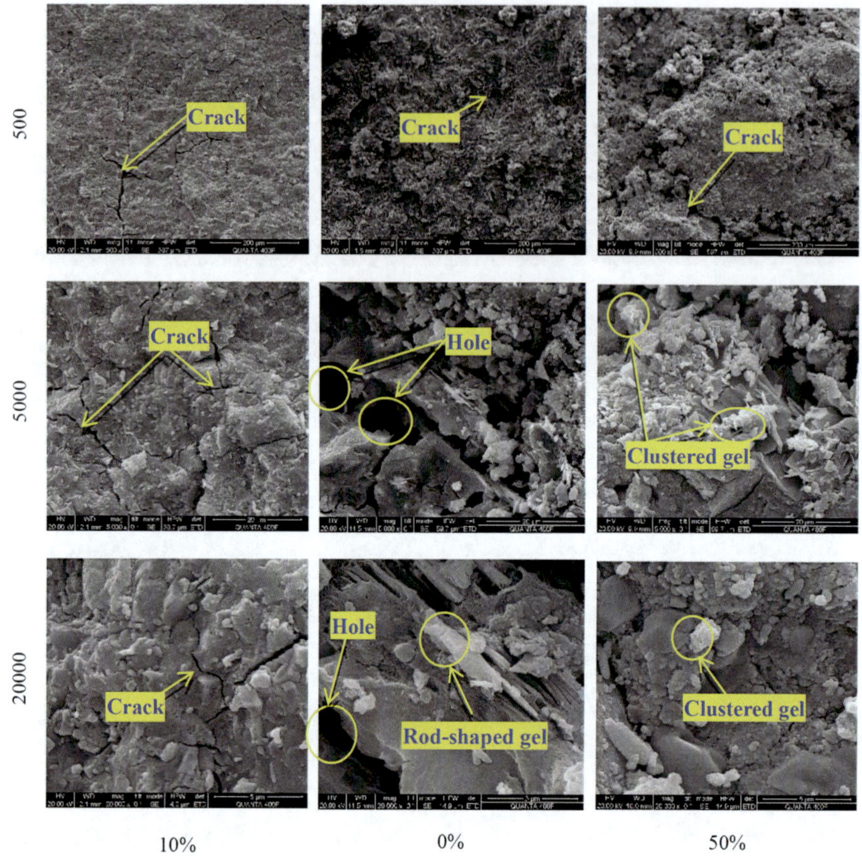

Fig. 4.18 Microstructural morphology of samples with different electrolytic manganese slag content

Table 4.1 Parameters of specimens with different water-reducing agent content

EMR content /%	W/C	Water-reducing agent content /%	Average pore size /nm	Apparent density /(g/mL)	Porosity /%	Permeability /md
30	0.4	0.4	52.51	2.3018	32.7662	1096.4894
30	0.4	0.6	52.83	2.294	33.3543	2465.07
30	0.4	0.8	60.9	2.3639	34.2727	277.5645

The influence of different superplasticizer dosages on the pore size distribution of the specimens is shown in Fig. 4.20. As the superplasticizer dosage increases, the porosity of the specimens gradually rises. When the superplasticizer content increases from 0.4% to 0.6%, the proportions of harmless pores and less harmful pores remain unchanged, while the proportion of harmful pores increases and that of multiharmful pores decreases. When the superplasticizer content increases from

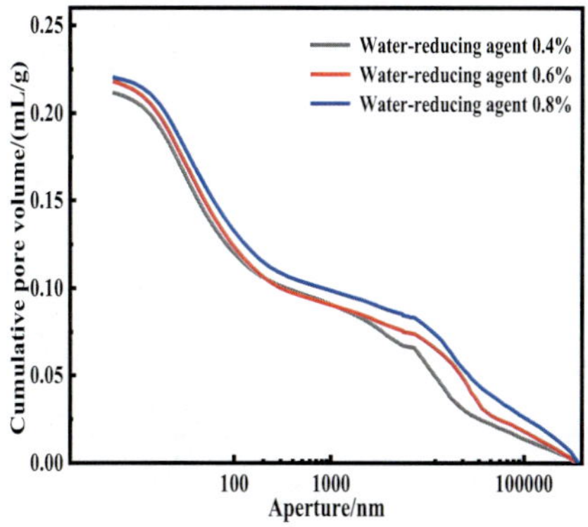

(a)

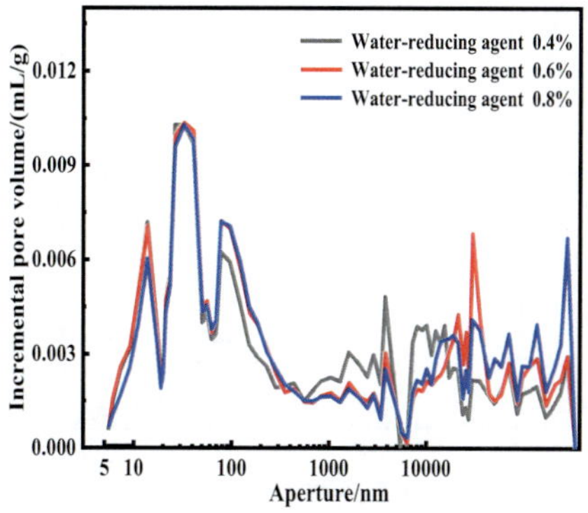

(b)

Fig. 4.19 Effect of different water-reducing agent dosages on the pore structure of test specimens

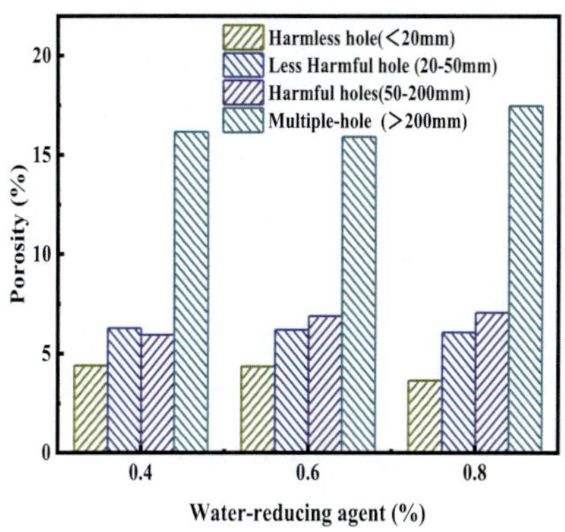

(a)

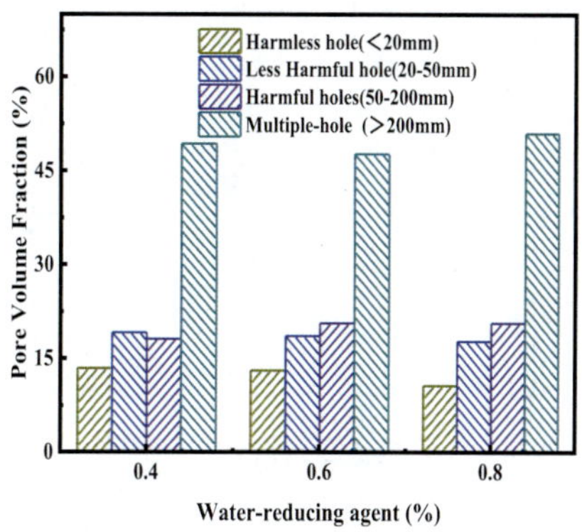

(b)

Fig. 4.20 Effect of different water-reducing agent dosages on pore distribution in test specimens

Table 4.2 Samples with different manganese slag content

EMR content /%	W/C	Water-reducing agent content /%	Average pore size /nm	Apparent density /(g/mL)	Porosity /%	Permeability / md
10	0.4	0.6	67.73	2.3431	29.9757	296.9146
30	0.4	0.6	52.83	2.294	33.3543	2465.07
50	0.4	0.6	140.07	1.8903	26.0682	478.5513

0.6% to 0.8%, the proportions of harmless pores and less harmful pores decrease, the proportion of harmful pores remains unchanged, and the proportion of multi-harmful pores increases.

The relevant parameters of specimens with different EMR contents are presented in Table 4.2. With a fixed superplasticizer dosage of 0.6%, as the EMR content

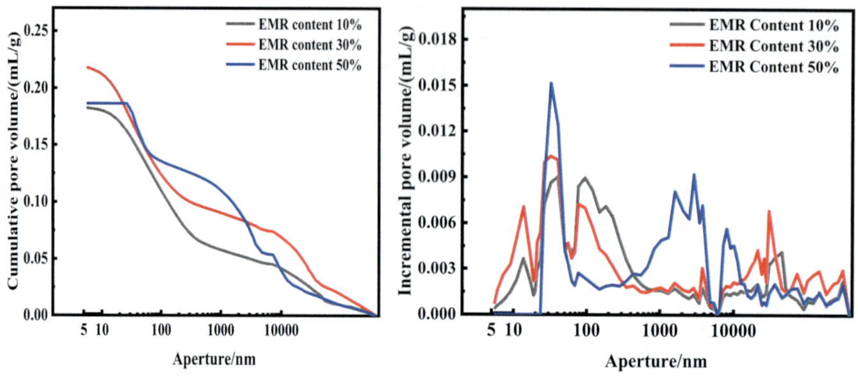

Fig. 4.21 Effect of different EMR dosages on the pore structure of test specimens

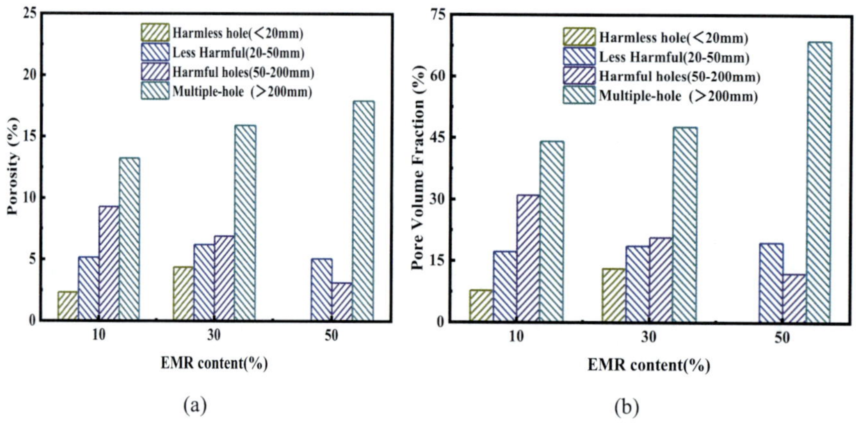

Fig. 4.22 Effect of different EMR dosages on the pore distribution of test specimens

increases, the porosity of the specimens first increases and then decreases; the average pore diameter first decreases and then increases; the apparent density shows a continuous decreasing trend; and the permeability first increases and then decreases.

The influence of different EMR contents on the pore structure of the specimens is shown in Fig. 4.21. Within the pore size range of 50–70 nm, the incremental pore volume increases with higher EMR content.

The influence of different EMR contents on the pore size distribution of the specimens is shown in Fig. 4.22. As the EMR content increases, the porosity of the specimens first rises and then declines. When the EMR content increases from 10% to 30%, the proportions of harmless pores and less harmful pores increase by 5% and 2%, respectively, while the proportion of harmful pores decreases by 10%, and that of multiharmful pores increases by 4%. Notably, at an EMR content of 50%, the proportion of harmless pores in the specimen is 0%. This may be attributed to experimental error or the possibility that pores smaller than 20 nm in the specimen are predominantly closed pores.

Alkali-Free Accelerator

The relevant parameters of specimens with different dosages of the alkali-free accelerator are listed in Table 4.3. As the accelerator dosage increases, both the porosity and permeability of the specimens first decrease and then increase; the average pore diameter first increases and then decreases; and the apparent density increases with higher accelerator dosage, though the extent of this increase is relatively minor.

The effect of different accelerator dosages on the pore distribution of specimens is shown in Fig. 4.23. As the accelerator dosage increases, the pore volume first decreases and then increases. When the accelerator dosage reaches 6%, the specimen exhibits the lowest porosity. As the rapid-setting agent dosage increases, the proportion of harmless pores rises from 4% to 24%, while the proportion of harmful pores decreases from 73% to 50%.

The influence of different dosages of the alkali-free accelerator on the pore size distribution of the specimens is shown in Fig. 4.24. As the accelerator dosage

Table 4.3 Different content of alkali-free accelerators

EMR content /%	W/C	Alkali-free accelerator content /%	Average pore size / nm	Apparent density /(g/mL)	Porosity /%	Permeability / md
30	0.4	4	40.34	2.0091	28.7036	248.0934
30	0.4	6	41.36	2.0441	27.1777	67.1066
30	0.4	8	35.07	2.0607	28.8985	758.0254

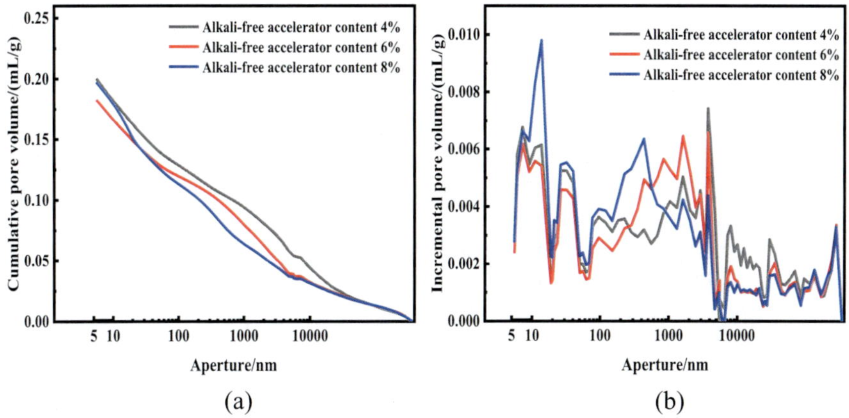

Fig. 4.23 Effect of different accelerator dosages on the pore structure of test specimens

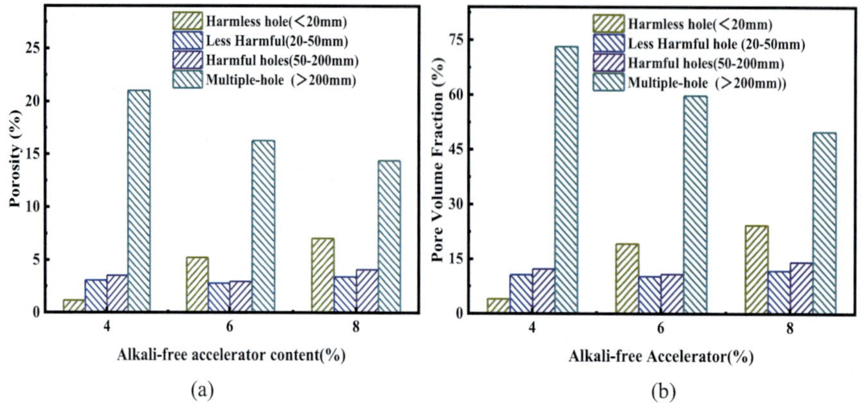

Fig. 4.24 Effect of different accelerator content on pore distribution in test specimens

Table 4.4 Parameters of samples with different manganese slag content

EMR content /%	W/C	Alkali-free accelerator content /%	Average pore size / nm	Apparent density /(g/mL)	Porosity /%	Permeability / md
10	0.4	6	50.4	2.0458	22.342	2124.05
30	0.4	6	41.36	2.0441	27.1777	67.1066
50	0.4	6	36.22	2.2522	42.1298	321.0605

increases, the porosity of the specimens first decreases and then increases, reaching its lowest value at an accelerator dosage of 6%. With increasing accelerator dosage, the proportion of harmless pores rises from 4% to 24%, while the proportion of multiharmful pores decreases from 73% to 50%.

Table 4.4 presents the relevant parameters of specimens with different EMR contents, at a fixed alkali-free accelerator dosage of 6%. As the EMR content increases, the porosity of the specimens gradually rises; the average pore diameter progressively decreases; while both the apparent density and permeability first decrease and then increase.

Figure 4.25 shows the influence of different EMR contents on the pore structure of the specimens. As the EMR content increases, the cumulative pore volume also continuously rises, with the extent of increase becoming more pronounced. Within the pore size range of 5–1000 nm, the incremental pore volume continuously increases with higher EMR content.

The influence of different EMR contents on the pore size distribution of the specimens is shown in Fig. 4.26. As the EMR content increases, the porosity of the specimens continuously rises, from 22.3% to 42.1%. The proportions of harmless pores, less harmful pores, and harmful pores all increase, while the proportion of multiharmful pores decreases.

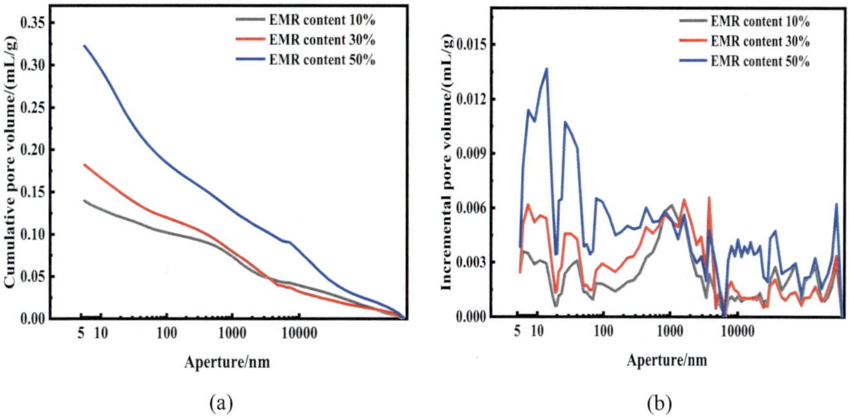

Fig. 4.25 Effect of different EMR dosages on the pore structure of test specimens

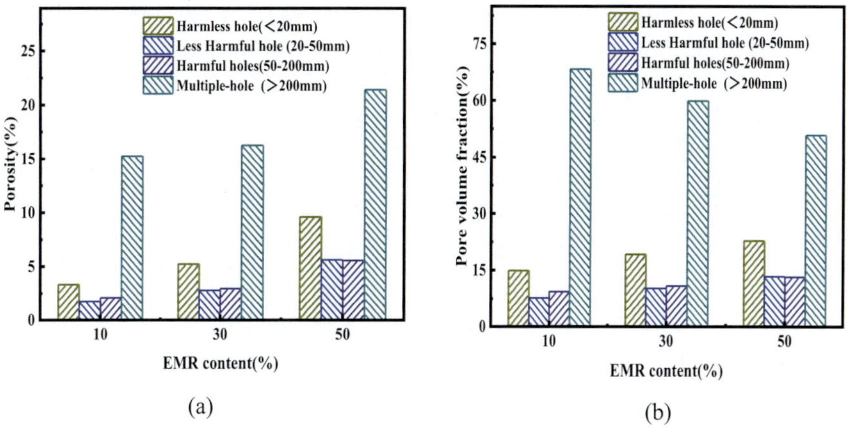

Fig. 4.26 Effect of different EMR dosages on pore distribution in specimens

4.3 Effect of Admixtures on the Performance of ESGM

4.3.1 Effect of Admixtures on the Fresh Properties of EFGM Grout

4.3.1.1 Flowability

Effect of Polycarboxylate Superplasticizer on the Flowability of ESGM Grout

Figure 4.27 illustrates the effect of different polycarboxylate superplasticizer dosages on the flowability of ESGM. The flowability increases with a higher superplasticizer dosage, although the extent of this increase is relatively limited. When the

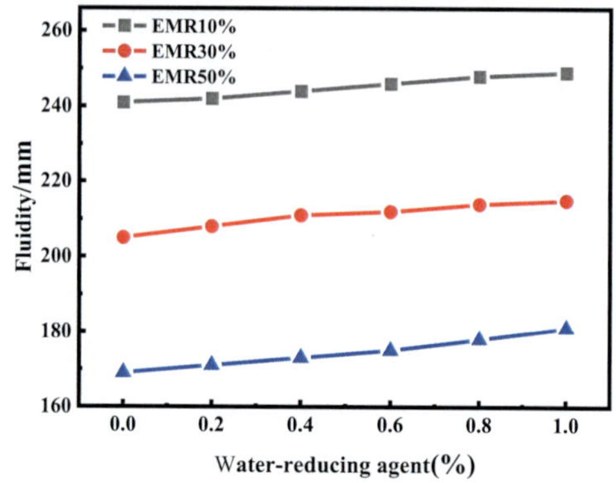

Fig. 4.27 Effect of polycarboxylic acid superplasticizer dosage on the flowability of ESGM

superplasticizer dosage changes from 0% to 1.0%, the flowability increases by 8 mm (a 3.3% increase) at an EMR content of 10%, by 10 mm (a 4.9% increase) at 30% EMR, and by 12 mm (a 7.1% increase) at 50% EMR. The polycarboxylate superplasticizer demonstrates a more pronounced improvement effect on grouts with higher EMR content. While the superplasticizer can enhance grout flowability to some extent, its overall influence is modest, which may be attributed to the relatively high water-to-binder ratio used. The addition of the superplasticizer reduces the attractive forces between grout particles, thereby lowering the yield stress. It also alters the nucleation and growth of ettringite, contributing to increased flowability. However, grouting materials typically employ a high water-to-binder ratio, and the influence of this ratio on grout properties far outweighs that of the superplasticizer.

Effect of Alkali-Free Accelerator on the Flowability of ESGM Grout

Figure 4.28 illustrates the effect of different dosages of the alkali-free accelerator on the flowability of ESGM grout. As the accelerator dosage increases, the flowability of the ESGM grout gradually decreases. When the accelerator dosage changes from 0% to 10%, the flowability decreases by 49 mm (a 20.3% reduction) at an EMR content of 10%, by 56 mm (a 27.3% reduction) at 30% EMR, and by 45 mm (a 26.6% reduction) at 50% EMR.

The alkali-free accelerator has a significant impact on grout flowability. Its primary component, aluminum sulfate, increases the concentration of aluminum ions and sulfate ions in the grout. These ions react rapidly with calcium ions to form ettringite, consuming calcium ions from the system and accelerating the dissolution of the raw materials. This process also consumes free water from the liquid phase, thereby reducing flowability. Furthermore, it is well known that a lower water-to-binder ratio decreases flowability. The addition of the accelerator effectively reduces the relative water-to-binder ratio, leading to a decrease in flowability.

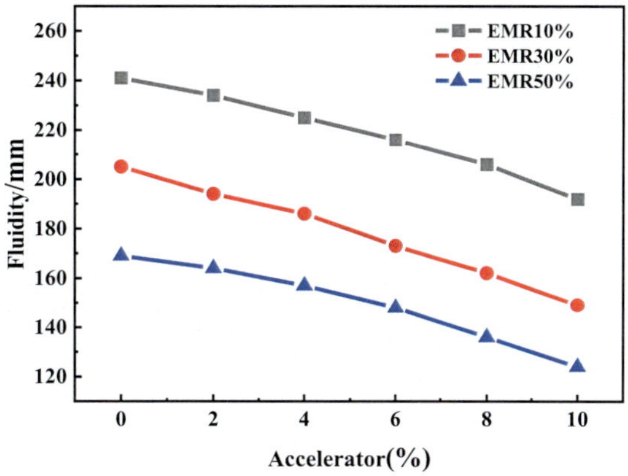

Fig. 4.28 Effect of alkali-free accelerator dosage on the flowability of ESGM

4.3.1.2 Initial and Final Setting Times

Effect of Polycarboxylate Superplasticizer on the Initial and Final Setting Times of ESGM Grout

Figure 4.29 illustrates the effect of different polycarboxylate superplasticizer dosages on the initial and final setting times of ESGM. As the superplasticizer dosage increases, both the initial and final setting times of the grout are progressively prolonged. When the superplasticizer dosage changes from 0% to 1.0%, at an EMR content of 10%, the initial setting time increases by 83 min (a 48.5% increase) and the final setting time by 100 min (a 53.7% increase). For a 30% EMR content, the initial and final setting times increase by 87 min (57.6%) and 107 min (63.7%), respectively. At a 50% EMR content, the increases are 83 min (70.3%) for the initial setting time and 91 min (64.5%) for the final setting time.

The superplasticizer significantly affects the setting time. It reduces the attractive forces between powder particles, decreasing inter-particle contact and thereby slowing down the degree of hydration. This retardation in the formation of hydration products leads to prolonged setting times. Furthermore, the polycarboxylate superplasticizer is acidic. Its introduction alters the alkaline environment of the grout, reducing its alkalinity. This weakens the alkali activation process necessary for forming cementitious products, further extending the setting time.

Effect of Alkali-Free Accelerator on the Initial and Final Setting Times of ESGM Grout

Figure 4.30 illustrates the effect of different dosages of the alkali-free accelerator on the initial and final setting times of ESGM. With an increase in the alkali-free accelerator dosage, both the initial and final setting times of the ESGM grout are significantly shortened. When the accelerator dosage changes from 0% to 1.0%, at an EMR content of 10%, the initial setting time is reduced by 128 min (a 74.9%

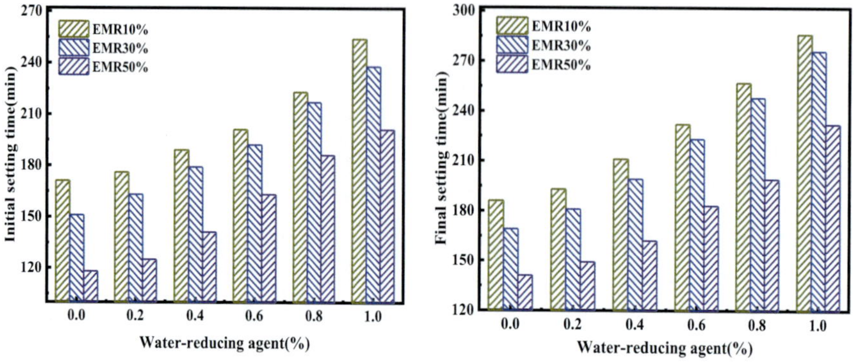

(a) Initial setting time (b) Final setting time

Fig. 4.29 Effect of polycarboxylic acid superplasticizer dosage on initial and final setting times of ESGM. (**a**) Initial setting time. (**b**) Final setting time

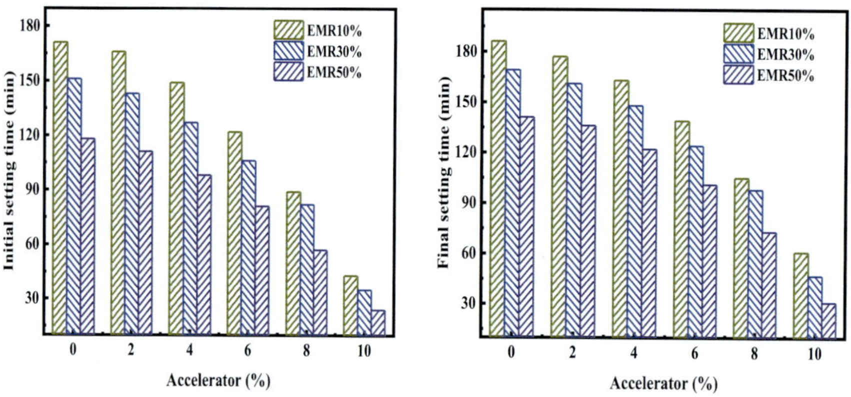

(a) Initial setting time (b) Final setting time

Fig. 4.30 Effect of alkali-free accelerator dosage on initial and final setting times of ESGM. (**a**) Initial setting time. (**b**) Final setting time

decrease) and the final setting time by 125 min (a 67.2% decrease). For a 30% EMR content, the reductions are 116 min (76.8%) for the initial setting time and 122 min (72.1%) for the final setting time. At a 50% EMR content, the decreases are 94 min (79.7%) for the initial setting time and 110 min (78.0%) for the final setting time.

The alkali-free accelerator substantially affects the setting time. On one hand, its introduction increases the concentration of aluminum ions and sulfate ions in the system, promoting the rapid formation of ettringite and C-(A)-S-H gel in the grout. This accelerates the hydration process, which consumes free water in the grout, shortens the hydration induction period, and thereby hastens grout setting. On the other hand, the addition of the accelerator effectively reduces the water-to-binder ratio. Particularly at a dosage of 10%, this contributes to a certain degree of setting time reduction.

4.3.2 *Effect of Admixtures on the Strength Characteristics of ESGM*

4.3.2.1 Effect of Polycarboxylate Superplasticizer on the Strength Characteristics of ESGM

Figure 4.31 illustrates the effect of different dosages of polycarboxylate superplasticizer on the compressive strength of ESGM. The compressive strength decreases with increasing superplasticizer content, indicating an adverse effect of the superplasticizer on strength development. When the superplasticizer dosage is increased from 0% to 1.0%, the 3-day, 14-day, and 28-day compressive strengths of specimens with 10% EMR content decrease by 34.8%, 40.9%, and 20.0% (or 16.4% for the 28-day strength under a specific reference), respectively. For specimens with 30% EMR, the reductions are 34.7%, 15.7%, and 18.7%, respectively. For those with 50% EMR, the decreases are 16.8%, 10.9%, and 12.2%, respectively.

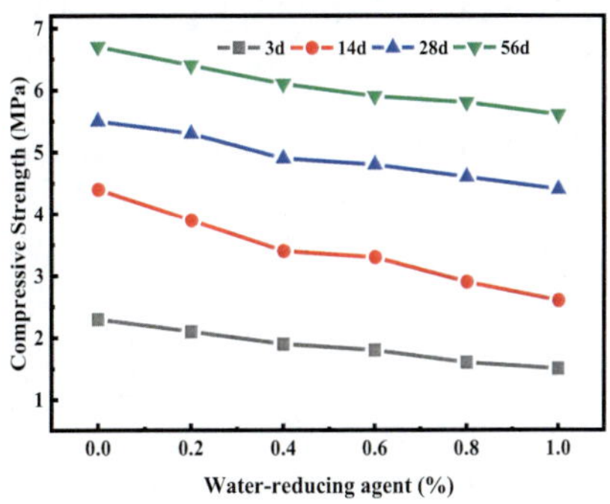

(a) EMR 10% compressive strength

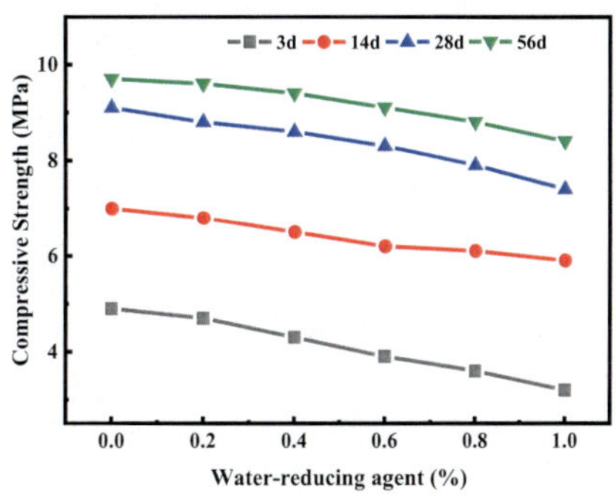

(b) EMR 30% compressive strength

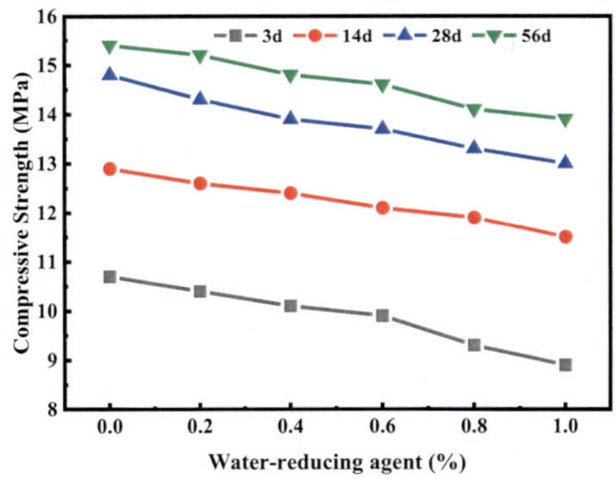

(c) EMR 50% compressive strength

Fig. 4.31 Effect of polycarboxylic acid superplasticizer on compressive strength of ESGM. (**a**) EMR 10% compressive strength. (**b**) EMR 30% compressive strength. (**c**) EMR 50% compressive strength

The superplasticizer disperses the powder particles through adsorption, reducing inter-particle contact. This inhibits the progress of the alkali-activated geopolymerization reaction, leading to a decrease in the amount of cementitious products formed and consequently reducing the specimen strength. Furthermore, the superplasticizer alters the nucleation and growth of ettringite, which may also negatively affect compressive strength. Additionally, the polycarboxylate superplasticizer is acidic and consumes part of the alkali in the system. As a result, some of the raw materials remain unactivated, reducing gel formation and further impairing compressive strength development.

4.3.2.2 Effect of Alkali-Free Accelerator on the Strength Characteristics of ESGM

Figure 4.32 illustrates the effect of different dosages of the alkali-free accelerator on the compressive strength of ESGM. When the EMR content is 10%, the compressive strength first increases and then decreases with increasing accelerator dosage, reaching its maximum at a dosage of 6%. Compared to specimens without the accelerator, the 3-day, 14-day, and 28-day compressive strengths are increased by 113.0%, 109.1%, and 86.3%, respectively. At an EMR content of 30%, the variation in compressive strength is similar to that at 10% EMR, with the maximum strength achieved at an accelerator dosage of 2%. The corresponding increases in 3-day, 14-day, and 28-day strengths are 14.3%, 8.6%, and 4.4%, respectively. When the EMR content is 50%, the accelerator has a negative impact on compressive strength; strength decreases with increasing accelerator dosage. At a dosage of 10%, the 3-day, 14-day, and 28-day compressive strengths are reduced by 21.5%, 15.5%, and 13.5%, respectively, relative to specimens without the accelerator.

The accelerator introduces aluminum ions and sulfate ions, promoting the formation and growth of ettringite as well as the substantial generation of C-(A)-S-H gel. Simultaneously, it consumes calcium ions in the system, accelerating early-age hydration. Ettringite and other gels interweave to form a dense three-dimensional network that provides strength. At lower EMR contents (e.g., 10%), the accelerator has a positive effect on ESGM, with an optimal dosage existing, and it significantly influences early-age strength. Furthermore, the incorporation of the accelerator reduces the water-to-binder ratio to some extent, which is beneficial for strength development. However, calcined EMR contains a considerable amount of calcium sulfate. At higher EMR contents, the sulfate ion concentration in the system is already elevated, and the addition of the accelerator further increases it. Excessive sulfate ions can combine with calcium ions and water to form gypsum, which undergoes volumetric expansion. This leads to increased internal microcracks in the specimen, thereby reducing its strength [32–34]. While the alkali-free accelerator speeds up grout hydration, excessively rapid hydration may not allow sufficient time for the hydration products to grow adequately. Consequently, they may fail to form tight, integrated connections and fully fill the pores, resulting in increased porosity and associated strength loss. Moreover, overly intense hydration can generate more microcracks, which is also detrimental to strength.

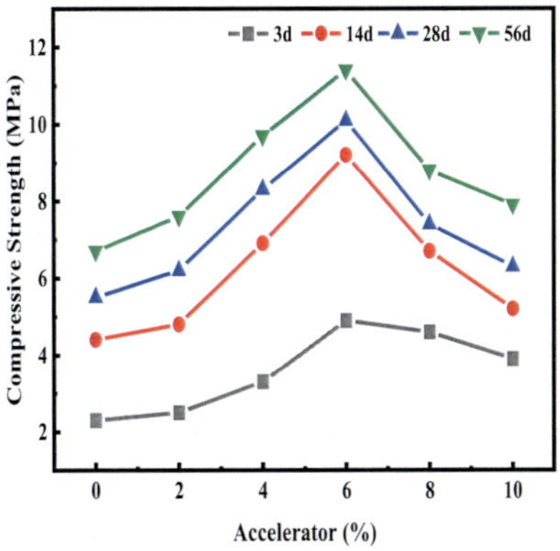

(a) EMR 10% compressive strength

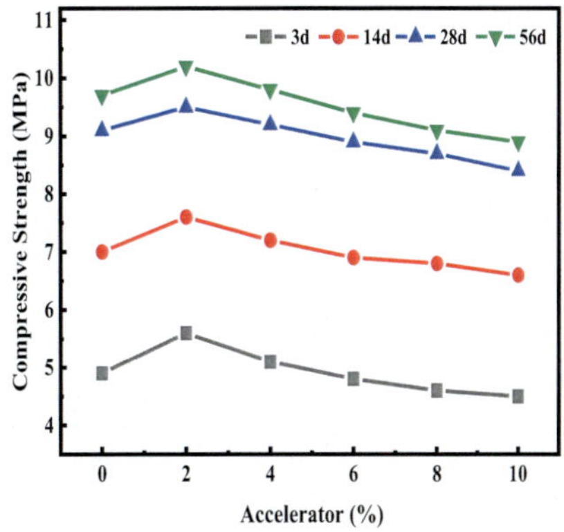

(b) EMR 30% compressive strength

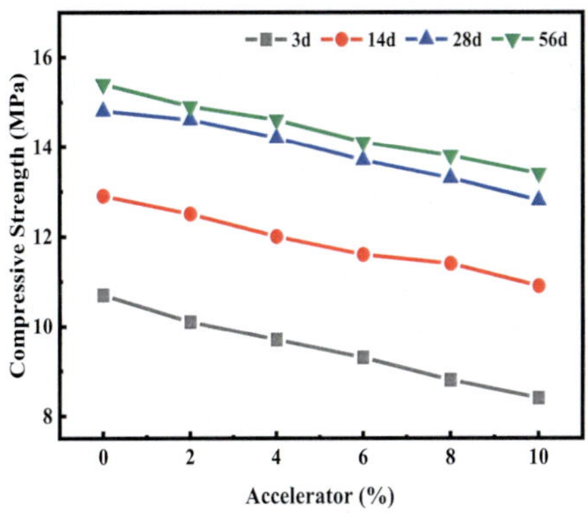

(c) EMR 50% compressive strength

Fig. 4.32 Effect of alkali-free accelerator on compressive strength of ESGM. (**a**) EMR 10% compressive strength. (**b**) EMR 30% compressive strength. (**c**) EMR 50% compressive strength

4.3.3 Effect of Admixtures on the Microstructural Characteristics of ESGM

4.3.3.1 XRD

The XRD patterns of ESGM specimens with a curing age of 28 days and an EMR content of 10%, under different superplasticizer dosages, are shown in Fig. 4.33. The hydration products formed in specimens with different superplasticizer contents are largely similar. As the superplasticizer dosage increases, the diffraction peaks of quartz, ettringite, calcite, and gypsum are enhanced. This is attributed to the reduced alkalinity of the system after superplasticizer addition, which diminishes the formation of C-(A)-S-H and N-A-S-H gels. Consequently, the consumption of CaO, reactive SiO_2, and Al_2O_3 is decreased. The unconsumed CaO combines with sulfate ions and carbon dioxide to form gypsum and calcite. The reduction in

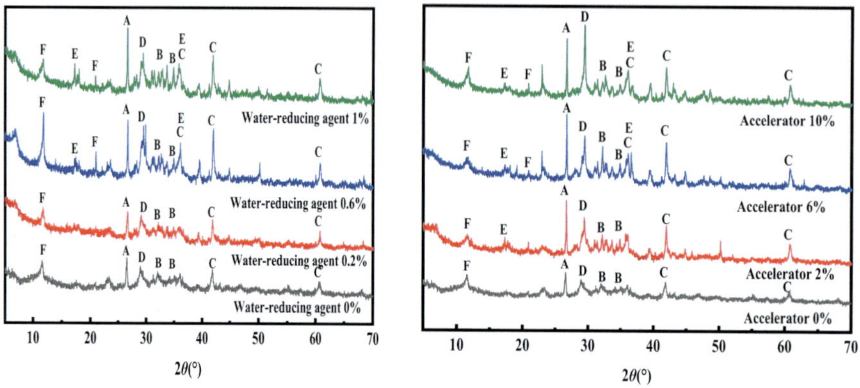

(a) Water-reducing agent (b) Accelerator

Fig. 4.33 28-day XRD patterns of ESGM specimens with admixtures (A: Quartz SiO_2, B: Calcium vanadate $Ca_6Al_2(SO_4)_3(OH)_{12}·26H_2O$, C: RO phase CaO-FeO-MnO-MgO solid solution, D: Calcite $CaCO_3$, E: Manganese oxide Mn_3O_4, MnO_2, F: Gypsum $CaSO_4·2H_2O$)

C-(A)-S-H and N-A-S-H gels leads to a decrease in the macroscopic strength of the specimens.

The XRD patterns of ESGM specimens with a curing age of 28 days and an EMR content of 10%, under different dosages of the alkali-free accelerator, are shown in Fig. 4.33. The intensity of the ettringite diffraction peaks continuously increases with higher accelerator dosage, indicating that the addition of the accelerator promotes the hydration reaction in the specimens, leading to the formation of more ettringite. The ettringite diffraction peaks are more pronounced in the specimen with a 6% accelerator content, while the gypsum diffraction peaks remain relatively weak. This suggests that more CaO and sulfate ions participate in the formation of ettringite, reducing gypsum formation, which is beneficial for strength development. Furthermore, at a low EMR content, the sulfate ion concentration in the system is limited, and the SS exhibits low reactivity while being present in a high proportion. Without the accelerator, the hydration reaction proceeds slowly, resulting in fewer cementitious products. With the addition of the accelerator, the increased sulfate and aluminum ions promote the rapid formation of ettringite, which quickly consumes calcium ions and accelerates the hydration rate. This indicates that the strength improvement in specimens with 10% EMR is primarily attributed to the accelerated formation and increased content of ettringite.

4.3.3.2 FTIR

Figure 4.34 presents the FTIR spectral analysis of ESGM specimens with a curing age of 28 days and an EMR content of 10%, under different admixture dosages. The vibrational absorption bands in the ranges of 3387–3470 cm^{-1} and 1648–1655 cm^{-1} are assigned to –O–H groups, associated with the presence of H_2O. The absorptions

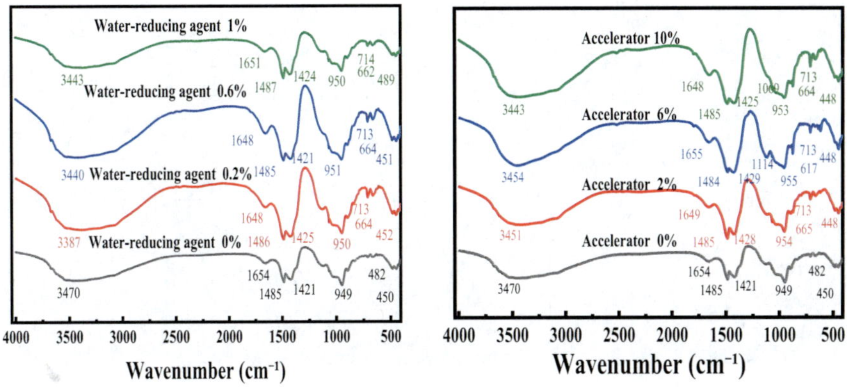

(a) Water-reducing agent (b) Accelerator

Fig. 4.34 28-day FTIR spectrum of ESGM specimen with admixture. (**a**) Water-reducing agent. (**b**) Accelerator

at 1421–1429 cm^{-1} and 1484–1487 cm^{-1} correspond to O–C–O groups, related to the presence of carbonates, primarily resulting from the formed calcite and other carbonate phases. The vibrational absorptions at 1069 cm^{-1} and 1114 cm^{-1} are attributed to Si–O–Si(Al) bonds. This absorption band arises from the formation of C-(A)-S-H and N-A-S-H gels during hydration. In Fig. 4.34a, the absorption peak in this region is not prominent, indicating the formation of fewer gels. In Fig. 4.34b, this absorption peak is stronger in specimens with accelerator dosages of 6% and 10%, suggesting a greater formation of such gels.

The vibrational absorption in the range of 949–955 cm^{-1} is assigned to Mn–O group vibrations, mainly originating from Mn_3O_4 and MnO_2 formed after the calcination of EMR. The absorptions at 713 cm^{-1}, 714 cm^{-1}, and 617–665 cm^{-1} are attributed to SO_4^{2-}, primarily associated with the formed ettringite. In Fig. 4.34b, the vibrational absorption in this region becomes stronger in specimens containing the accelerator, indicating that the accelerator addition promotes ettringite formation. The vibrational absorption in the range of 448–489 cm^{-1} corresponds to the deformation vibration of Si–O–Si, which is related to SiO_2, indicating the presence of non-reactive SiO_2.

SEM-DES

Figure 4.35 shows the micromorphology of specimens with an EMR content of 10%, an SS content of 60%, and a curing age of 28 days under different superplasticizer dosages. As the superplasticizer dosage increases, the specimens gradually become looser. In the absence of the superplasticizer (Fig. 4.35a), the specimen contains a few relatively large pores, but the gels are bonded into a relatively dense and integrated structure, providing strength. At a superplasticizer dosage of 0.2%

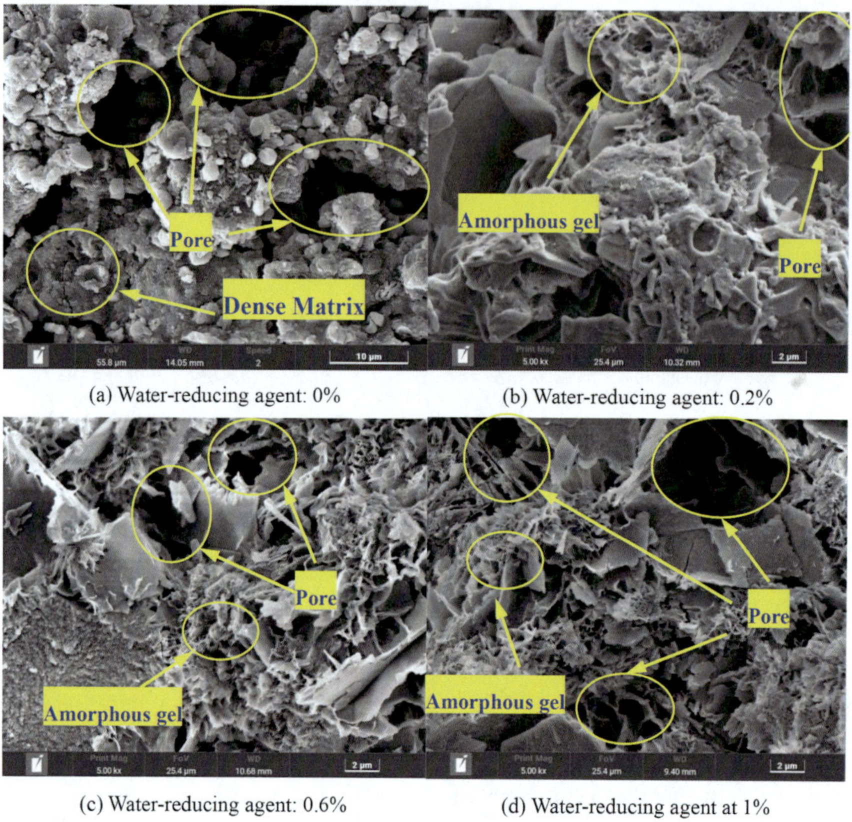

Fig. 4.35 28-day SEM image of EMR (10%) + SS (60%) specimen under polycarboxylic acid superplasticizer influence. (**a**) Water-reducing agent: 0%. (**b**) Water-reducing agent: 0.2%. (**c**) Water-reducing agent: 0.6%. (**d**) Water-reducing agent at 1%

(Fig. 4.35b), no large pores are observed in the specimen; however, numerous small pores exist between the amorphous gels, resulting in a lower density of the formed gel network. At superplasticizer dosages of 0.6% and 1.0% (Fig. 4.35c, d), not only do the small pores between amorphous gels increase, but larger pores also appear, further reducing the compactness and strength of the specimens. This may result from the interaction between the superplasticizer and ettringite crystals, where the superplasticizer slows down or inhibits the growth of ettringite. The SEM results indicate that the addition of the polycarboxylate superplasticizer increases the porosity in the specimens, thereby reducing their strength.

Figure 4.36 shows the micromorphology of specimens with an EMR content of 10%, an SS content of 60%, and a curing age of 28 days under different dosages of the alkali-free accelerator. Compared to the specimen without the accelerator (Fig. 4.36a), when the accelerator dosage is 2% (Fig. 4.36b), some of the larger pores in the specimen are filled with amorphous gels and ettringite, resulting in smaller pores, a denser structure, and improved specimen strength. At an

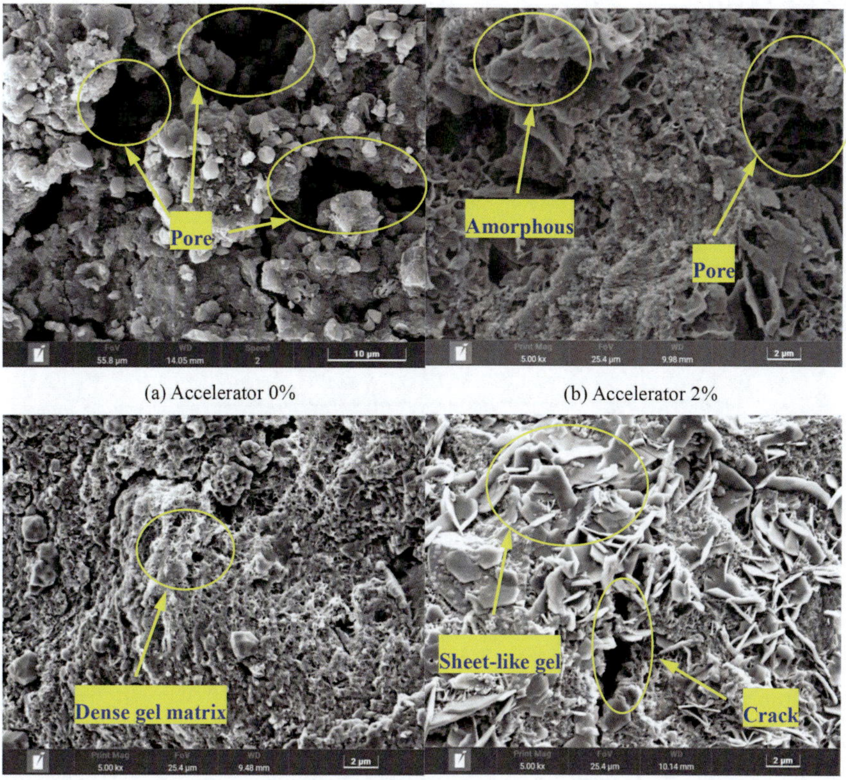

(a) Accelerator 0%

(b) Accelerator 2%

(c) Accelerator 6%

(d) Accelerator 10%

Fig. 4.36 28-day SEM image of EMR (10%) + SS (60%) specimen under the influence of alkali-free accelerator. (**a**) Accelerator 0%. (**b**) Accelerator 2%. (**c**) Accelerator 6%. (**d**) Accelerator 10%

accelerator dosage of 6% (Fig. 4.36c), the gels and ettringite in the specimen are interbonded, forming a dense microstructure with high integrity, where few or no obvious pores are observed. When the accelerator dosage reaches 10% (Fig. 4.36d), a large amount of flaky gels (likely $Ca(OH)_2$) and ettringite are interlocked in the specimen, resulting in a relatively dense microstructure and good overall integrity. Additionally, isolated larger cracks are present in the specimen, which may be caused by the volumetric expansion of gypsum formed from excessive SO_4^{2-} and Ca^{2+} when the accelerator content is too high. The SEM results indicate that the addition of the alkali-free accelerator promotes the formation of more gels and ettringite, which fill the pores and enhance the compactness of the specimens, thereby improving their strength. However, excessive alkali-free accelerator leads to the formation of a significant amount of expansive gypsum, causing specimen cracking and a consequent reduction in mechanical strength with further increases in accelerator dosage.

4.3.3.3 MIP

Figure 4.37 and Table 4.5 present the pore size distribution test results of ESGM specimens with an EMR content of 10% and a curing age of 28 days, incorporating different admixtures. As shown in Fig. 4.37a, compared to the specimen without admixtures, the cumulative pore volume distribution curve of the specimen with 0.6% superplasticizer shifts toward larger pores, while that of the specimen with 6% accelerator shifts significantly toward smaller pores. This indicates that the addition of the accelerator is beneficial for reducing the distribution of cumulative pore volume, whereas the superplasticizer has the opposite effect. In Fig. 4.37b, the incremental pore volume distribution curve of the specimen with 6% accelerator shows

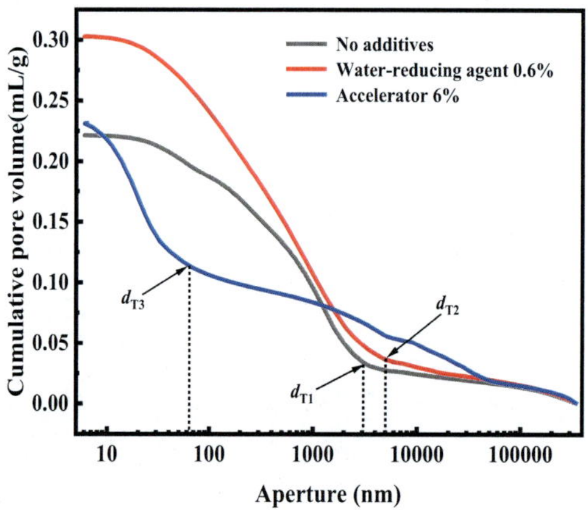

(a) Cumulative pore volume distribution curve

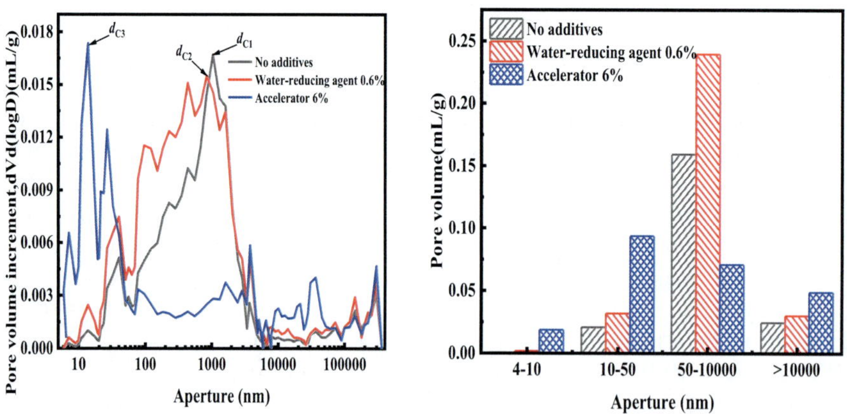

(b) Pore volume increment distribution curve (c) Pore Volume Distribution Statistics

Fig. 4.37 28-day MIP test results for ESGM specimens under the influence of admixtures. (**a**) Cumulative pore volume distribution curve. (**b**) Pore volume increment distribution curve. (**c**) Pore Volume Distribution Statistics

Table 4.5 Pore structure parameters of ESGM specimens with admixtures at 28 days

Sample	Threshold aperture d_T(nm)	Critical Aperture d_C(nm)	Average pore size (nm)	Apparent density (g/mL)	Cumulative pore volume (mL/g)	Porosity (%)
No additives	2920.94	1039.67	173.05	2.7474	0.221	37.82
Water-reducing agent 0.6%	4817.69	835.69	133.09	2.6894	0.303	44.85
Accelerator 6%	62.51	13.73	35.52	2.6231	0.231	37.72

a distinct leftward shift relative to the specimen without admixtures. The specimen with 0.6% superplasticizer also shifts slightly to the left, but the incremental pore volume corresponding to small pores increases substantially, and the area enclosed by its curve is noticeably larger.

In Fig. 4.37c, compared to the specimen without admixtures, the specimen with 6% accelerator exhibits a greater volume of pores smaller than 50 nm, while the pore volume in the range of 50–10,000 nm is significantly reduced. In contrast, the specimen with 0.6% superplasticizer shows a markedly larger pore volume within the 50–10,000 nm range.

As listed in Table 4.5, compared to the specimen without admixtures, the specimen with 6% accelerator exhibits reductions in the threshold pore diameter (d_T), critical pore diameter (d_C), average pore diameter, and porosity, with values of 62.51 nm, 13.73 nm, 35.52 nm, and 37.72%, respectively. In contrast, the specimen with 0.6% superplasticizer shows increases in the threshold pore diameter (d_T), average pore diameter, cumulative pore volume, and porosity, with values of 4817.69 nm, 133.09 nm, 0.303 mL/g, and 44.85%, respectively. These results indicate that at an EMR content of 10%, the incorporation of 6% accelerator helps reduce porosity and enhance macroscopic strength, while the addition of 0.6% superplasticizer increases porosity and reduces macroscopic strength.

4.4 Effect of Admixtures on the Performance of ECGM

4.4.1 Effect of Admixtures on the Fresh Properties of ECGM Grout

4.4.1.1 Flowability

Effect of Polycarboxylate Superplasticizer on the Flowability of ECGM Grout

Figure 4.38 illustrates the effect of different polycarboxylate superplasticizer dosages on the flowability of ECGM grout. The incorporation of the polycarboxylate superplasticizer plays a significant role in improving the flowability of ECGM

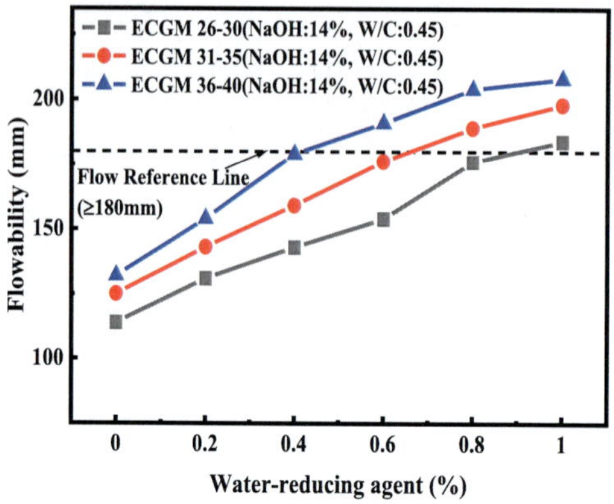

Fig. 4.38 Effect of polycarboxylic acid superplasticizer dosage on ECGM flow

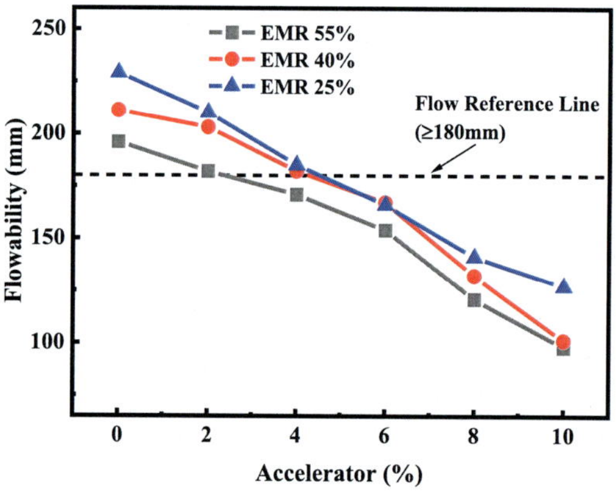

Fig. 4.39 Effect of alkali-free accelerator dosage on ECGM flowability

grout. As the superplasticizer dosage increases, the flowability of ECGM rises. At lower superplasticizer dosages (≤0.4%), the improvement in grout fluidity is more pronounced, indicating that a low dosage of the superplasticizer is relatively effective. However, at higher dosages, although the flowability increases to some extent, the rate of improvement becomes noticeably smaller, suggesting diminished water-reducing efficiency. This is because, with increasing superplasticizer dosage, the surfaces of particles within the reaction system become increasingly occupied by the adsorbed superplasticizer molecules [35], thereby reducing the effectiveness of further water reduction.

Effect of Alkali-Free Accelerator on the Flowability of EFGM Grout

Figure 4.39 illustrates the effect of different dosages of the alkali-free accelerator on the flowability of ECGM grout. As the dosage of the alkali-free accelerator increases, the flowability of the ECGM grout decreases. In contrast to the control group without the accelerator, the flowability of the test groups with accelerator gradually declines. Although the addition of the accelerator can significantly shorten the setting time and achieve early strength enhancement, it also reduces the fluidity of the grout and the later-age strength of the grouting material [36].

4.4.1.2 Initial and Final Setting Times

Effect of Polycarboxylate Superplasticizer on the Setting Times of ECGM

Figure 4.40 illustrates the effect of different polycarboxylate superplasticizer dosages on the initial and final setting times of ECGM grout. As the superplasticizer dosage increases, both the initial and final setting times of the grout first decrease and then increase. For all three mix proportions, the setting times of ECGM grout reach their minimum at a superplasticizer dosage of 0.6%. In the test group with 55% EMR content, the initial and final setting times are reduced by 23 min and 21 min, respectively, compared to the control group. For the group with 40% EMR content, the reductions are 24 min and 35 min, respectively, while for the group with 25% EMR content, they are 18 min and 14 min, respectively.

As shown in the figure, when the superplasticizer dosage exceeds 0.8%, the variation in setting times becomes less pronounced. Notably, at a dosage of 1.0%, the setting times of ECGM grout for all three mix proportions are longer than those of the control group. This indicates that excessive superplasticizer retards the setting of ECGM grout, primarily because an overdose forms an adsorption film on the surface of cementitious particles, which hinders the hydration process and prolongs the setting time [8].

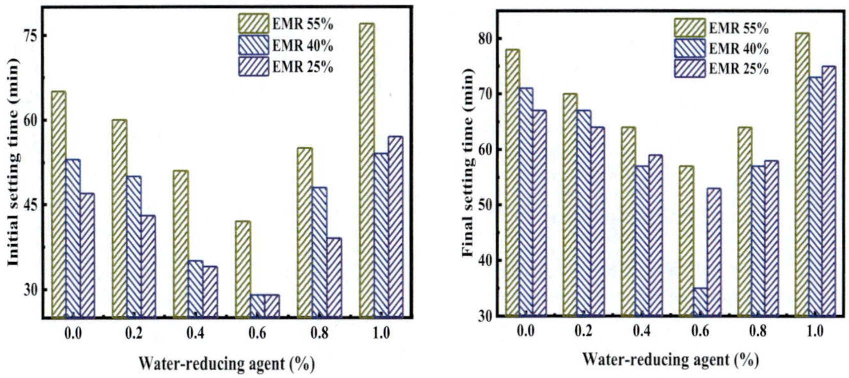

(a) Initial setting time (b) Final setting time

Fig. 4.40 Effect of polycarboxylic acid superplasticizer dosage on initial and final setting times of ECGM. (**a**) Initial setting time. (**b**) Final setting time

Effect of Alkali-Free Accelerator on the Setting Times of ECGM

Figure 4.41 illustrates the effect of different dosages of the alkali-free accelerator on the initial and final setting times of ECGM grout. As the dosage of the alkali-free accelerator increases, both the initial and final setting times of the ECGM grout first decrease and then increase, reaching their minimum at an accelerator dosage of 4%. For the test group with 55% EMR content, the initial and final setting times are shortened by 37 min and 29 min, respectively, compared to the group without the accelerator. In the group with 40% EMR, the reductions are 41 min and 30 min, and for the group with 25% EMR, they are 41 min and 39 min, respectively.

However, when the accelerator dosage exceeds 6%, the setting times begin to increase significantly. This effect is particularly pronounced at a dosage of 10%, where the setting times are substantially prolonged. The strong alkalinity of the grout promotes the conversion of Al^{3+} to $[Al(OH)_4]^-$, which reacts with Ca^{2+} and SO_4^{2-} in the system to form gel-like crystals [176], contributing to the shortening of the setting time. In summary, an appropriate dosage of the accelerator can accelerate the setting and hardening of ECGM. However, when the dosage exceeds 4%, the setting time tends to increase, and beyond 8%, it no longer exhibits a shortening effect on the setting time.

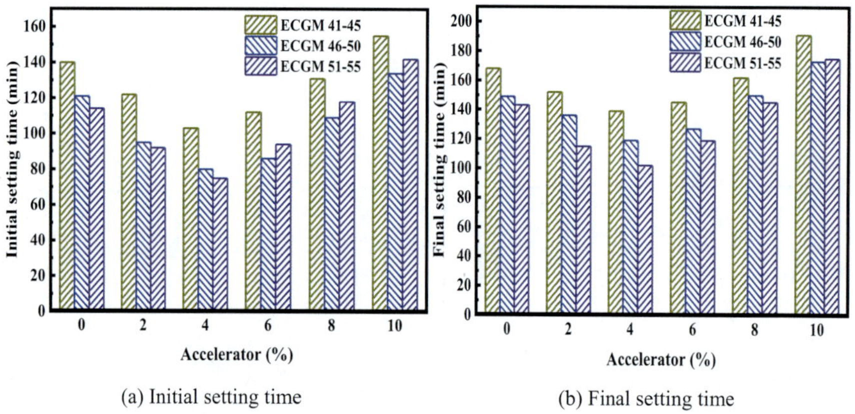

Fig. 4.41 Effect of alkali-free accelerator dosage on initial and final setting times of ECGM. (**a**) Initial setting time. (**b**) Final setting time

4.4.2 *Effect of Admixtures on the Strength Characteristics of ECGM*

4.4.2.1 The Influence of Polycarboxylate Superplasticizer on the Strength Characteristics of ECGM

Figure 4.42 illustrates the effect of different polycarboxylate superplasticizer dosages on the compressive strength of ECGM. The compressive strength progressively decreases with increasing superplasticizer dosage, indicating that the addition of the polycarboxylate superplasticizer adversely affects the compressive strength of ECGM. Moreover, the extent of this negative influence is largely consistent across all tested mix proportions. At a dosage of 1.0%, the 28-day compressive strength of ECGM specimens under all three mix proportions drops sharply.

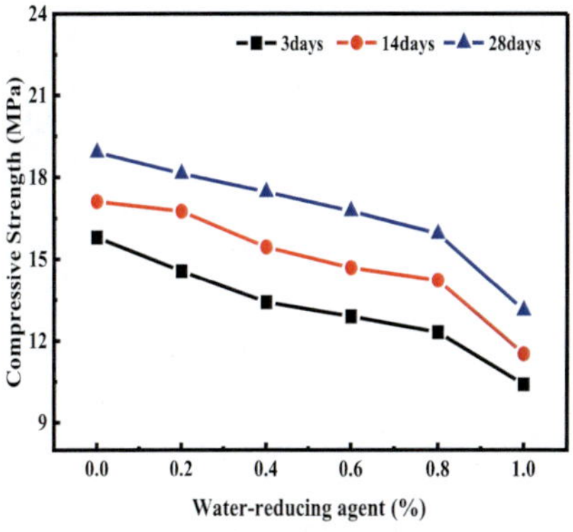

(a) EMR 25% compressive strength

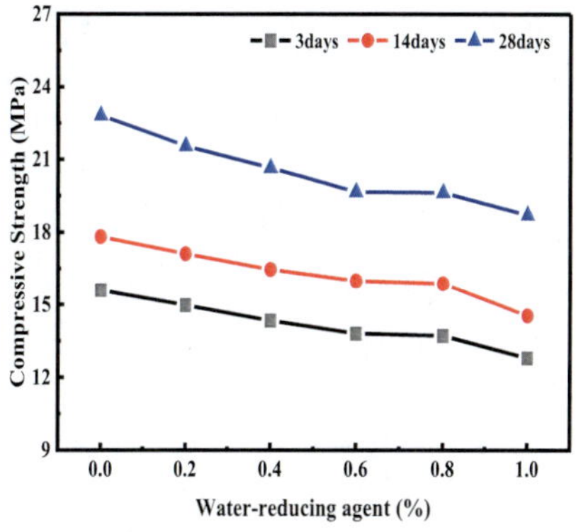

(b) EMR 40% compressive strength

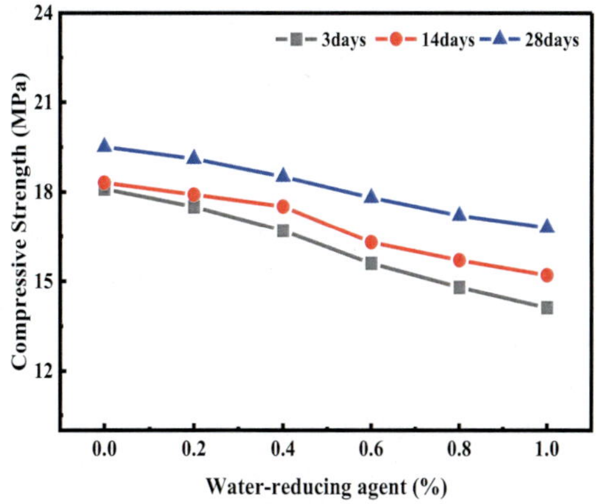

(c) EMR 55% compressive strength

Fig. 4.42 Effect of Polycarboxylic Acid Superplasticizer Dosage on ECGM Compressive Strength. (**a**) EMR 25% compressive strength. (**b**) EMR 40% compressive strength. (**c**) EMR 55% compressive strength

The polycarboxylate superplasticizer is an acidic high-range water reducer. In an alkaline environment, its stability may be compromised, and the acidic nature of the admixture can also reduce the alkalinity required for geopolymerization, thereby negatively impacting compressive strength. Additionally, although the polycarboxylate superplasticizer effectively enhances grout fluidity and shortens hydration time, it exerts a detrimental effect on the compressive strength of all test groups, which aligns with the findings reported by Sheng Yuhang [37].

4.4.2.2 Effect of Alkali-Free Accelerator on the Strength Characteristics of ECGM

Figure 4.43 illustrates the effect of different dosages of the alkali-free accelerator on the compressive strength of ECGM. The compressive strength of the specimens first increases and then decreases with increasing accelerator dosage, indicating that a certain amount of accelerator enhances the compressive strength of ECGM. For the test group with 55% EMR content, at an accelerator dosage of 2%, the 3-day, 14-day, and 28-day compressive strengths increased by 16.1%, 22.4%, and 28.0%, respectively, compared to the specimens without the accelerator. In the group with 40% EMR content, at a dosage of 4%, the strengths increased by 26.8%, 33.5%, and 35.6%, respectively. For the group with 25% EMR content, at a dosage of 6%, the increases were 27.4%, 38.7%, and 42.2%, respectively.

However, when the accelerator dosage exceeds a certain threshold, the compressive strength of all test groups declines with further increases in dosage, and the reduction becomes more pronounced at higher dosages. Furthermore, the accelerator showed the most significant improvement in compressive strength for the group with 25% EMR content, followed by the group with 40% EMR, and the least for the group with 55% EMR. This may be attributed to a better synergistic hydration effect between the ECGM mix proportion (with 25% EMR) and the alkali-free accelerator, promoting the formation of more cementitious products and thereby enhancing the compressive strength of the specimens [38, 39].

4.4.3 Effect of Admixtures on the Microstructural Characteristics of ECGM

4.4.3.1 XRD

Figure 4.44 shows the XRD patterns of ECGM specimens with three different mix proportions, both with and without the superplasticizer, at a curing age of 28 days. The results indicate that the primary cementitious hydration products in ECGM containing the superplasticizer are geopolymer-type gels, as indicated by the diffuse hump between 20° and 40°. A comparison with the control groups reveals that the

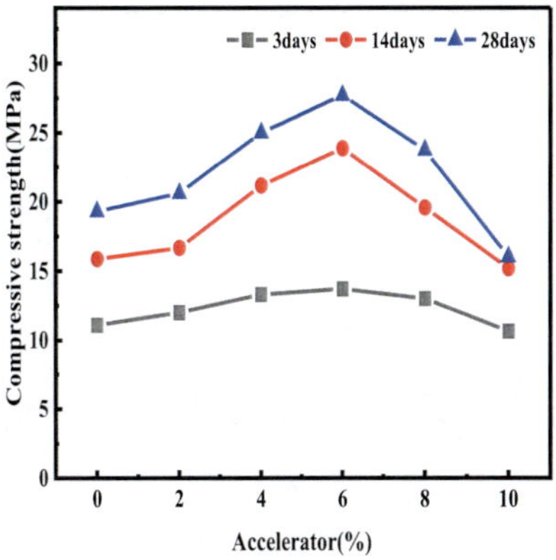

(a) EMR 25% compressive strength

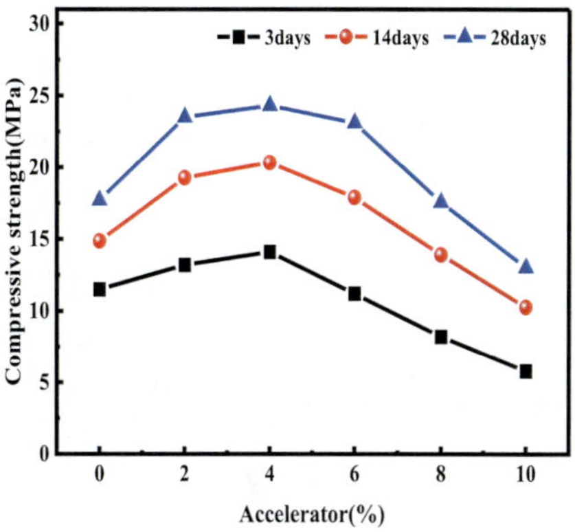

(b) EMR 40% compressive strength

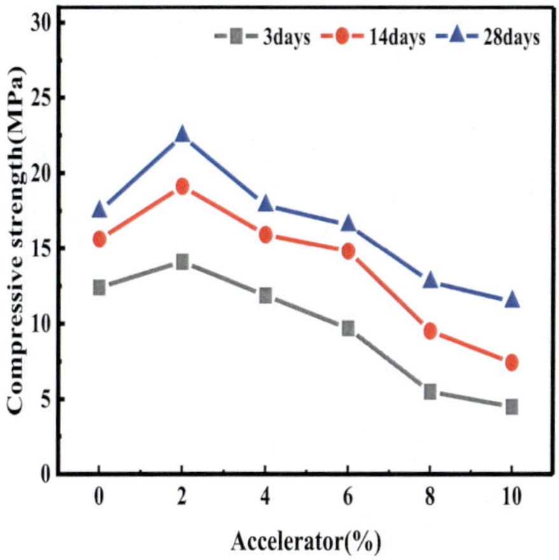

(c) EMR 55% compressive strength

Fig. 4.43 Effect of alkali-free accelerator dosage on ECGM compressive strength. (**a**) EMR 25% compressive strength. (**b**) EMR 40% compressive strength. (**c**) EMR 55% compressive strength

diffraction peaks corresponding to gel products are less intense in all test groups after the addition of the superplasticizer.

The diffraction patterns of ECGM with different mix proportions exhibit distinct peak characteristics. The specimen with 25% EMR content shows the largest hump area, suggesting the highest quantity of gel products formed, followed by the group with 40% EMR content, and then the group with 55% EMR content. With increasing superplasticizer dosage, phases such as ferro-aluminian gedrite begin to appear within the geopolymer gel matrix [40]. This occurs because the added superplasticizer alters the surface charge distribution of particles. Although this influence can reduce flocculation between particles to some extent, the polycarboxylate

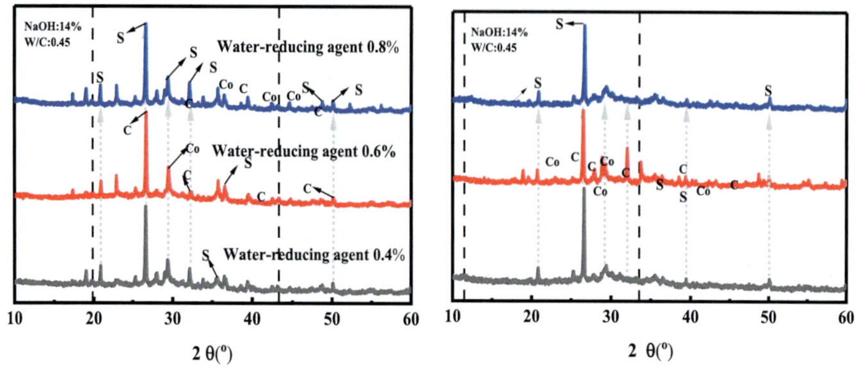

(a) Water-reducing agent (b) Control group

Fig. 4.44 28-day XRD patterns of ECGM specimens with different water-reducing agent dosages (S: SiO_2, C: $Ca_{1.5}SiO_{3.5}{\cdot}xH_2O$, Co: $CaCO_3$)

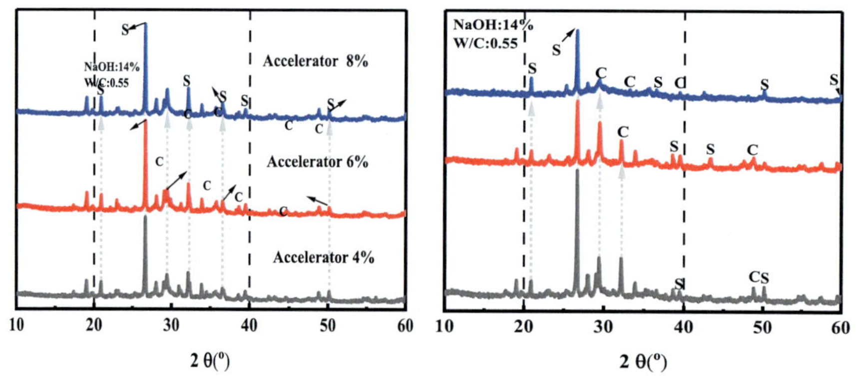

(c) Accelerator (d) Control group

Fig. 4.45 28-day XRD patterns of ECGM specimens with different alkali-free accelerator dosages (S: SiO_2, C: $Ca_{1.5}SiO_{3.5}{\cdot}xH_2O$, Co: $CaCO_3$)

superplasticizer primarily inhibits the formation of gel products, leading to a decrease in compressive strength.

Figure 4.45 presents the XRD patterns of ECGM specimens with three different mix proportions, both with and without the alkali-free accelerator, at a curing age of 28 days. The XRD diffraction peaks of the ECGM groups containing the accelerator exhibit distinct patterns. Compared to the control group without the accelerator, the addition of the accelerator results in more pronounced diffraction features corresponding to cementitious products in all test groups, with the effect being most significant in the group with 25% EMR content.

The calcium silicate hydrate (C-S-H) gel found in the hydration products of the ECGM test groups tends to form more readily when the Mg content in the system is low and is a common hydration gel product in alkali-activated cementitious materials [41]. Zhang Zhengan [42] observed that specimens with an alkali-free accelerator (6%–8%) already exhibited distinct diffraction peaks for geopolymer gel products within 6 min of the hydration reaction. Numerous studies have also indicated that grouts incorporating an alkali-free accelerator produce a substantial amount of short-columnar gel products within the first 10 min of hydration [43, 44]. Within 1 h, these gel products grow and interconnect, with C-S-H gels interpenetrating the network voids, demonstrating that the participation of the accelerator promotes the early formation of extensive geopolymer gel crystals, contributing to the early-age strength of the specimens.

4.4.3.2 SEM-EDS

Figure 4.46 presents the SEM micrographs of specimens with three different mix proportions, both with and without the addition of 0.4% superplasticizer, at a curing age of 28 days. As shown in Fig. 4.46a, b, the matrix of the ECGM 26–30 test group with superplasticizer contains more voids and cracks, and the amount of gel products generated by the hydration reaction is less compared to the group without superplasticizer.

From Fig. 4.46c, it can be observed that the specimens of the ECGM 31–35 test group with superplasticizer are denser than those of the ECGM 26–30 group, with fewer pores and voids. The amount of geopolymer gel formed is also greater than in the ECGM 26–30 group. However, compared to the ECGM 31–35 group without superplasticizer, the specimens with superplasticizer produce less geopolymer gel. Similarly, the ECGM 36–40 test group exhibits comparable characteristics—the addition of superplasticizer also negatively affects the microstructure. Nevertheless, as the EMR content decreases (i.e., the CSS content increases), the adverse effect of the superplasticizer on the microstructure diminishes. Among them, the ECGM 36–40 group shows the densest microstructure and the highest amount of geopolymer gel formed.

Furthermore, microscopic analysis indicates that the primary hydration products in the hardened specimens with superplasticizer are mainly N-C-S-A-H gel and C-S-A-H gel. These observations are consistent with the strength test results.

Figure 4.47 shows the EDS analysis results of the gel products from specimens of the ECGM 26–30 and ECGM 36–40 test groups, with a superplasticizer dosage of 0.4% and a curing age of 28 days. The elemental compositions at each spectral point in Fig. 4.47 are listed in Table 4.6. A comparison between Fig. 4.46b, d reveals that, under both mix proportions with the added superplasticizer, the gel products in the specimens consist of the same elements as those in Fig. 4.46b, d, although their concentrations differ. This discrepancy may be attributed to the varying effectiveness of the superplasticizer across different material mix proportions. Within the hardened specimens of the ECGM 36–40 group, which has a higher CSS content, a greater amount of hydrated cementitious products can form, which is the primary reason for its higher strength. The EDS results indicate that the CSS content significantly influences the formation of cementitious products in hardened specimens

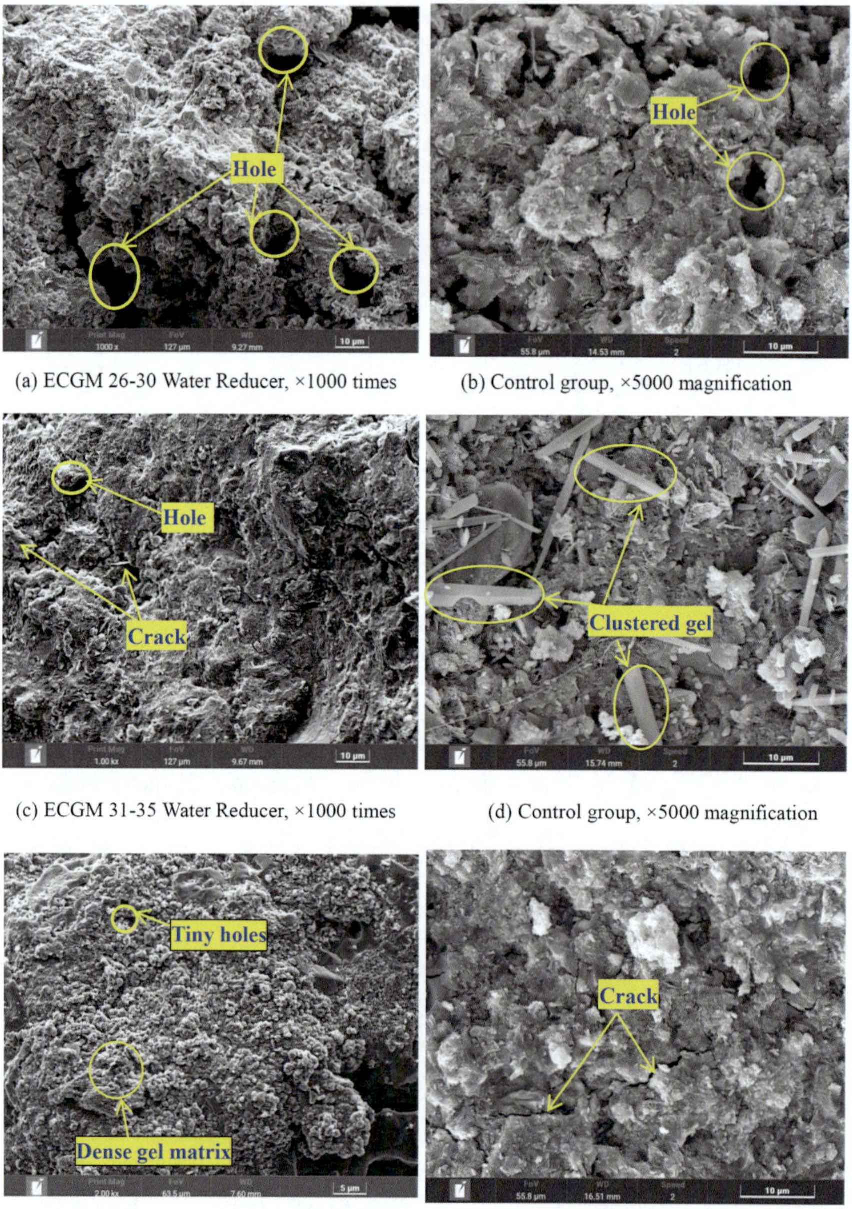

(a) ECGM 26-30 Water Reducer, ×1000 times

(b) Control group, ×5000 magnification

(c) ECGM 31-35 Water Reducer, ×1000 times

(d) Control group, ×5000 magnification

(e) ECGM 36-40 Water Reducer, ×2000 times

(f) Control group, ×5000 magnification

Fig. 4.46 SEM images of hardened specimens with different material ratios incorporating water-reducing agents. (**a**) ECGM 26–30 Water Reducer, ×1000 times. (**b**) Control group, ×5000 magnification. (**c**) ECGM 31–35 Water Reducer, ×1000 times. (**d**) Control group, ×5000 magnification. (**e**) ECGM 36–40 Water Reducer, ×2000 times. (**f**) Control group, ×5000 magnification

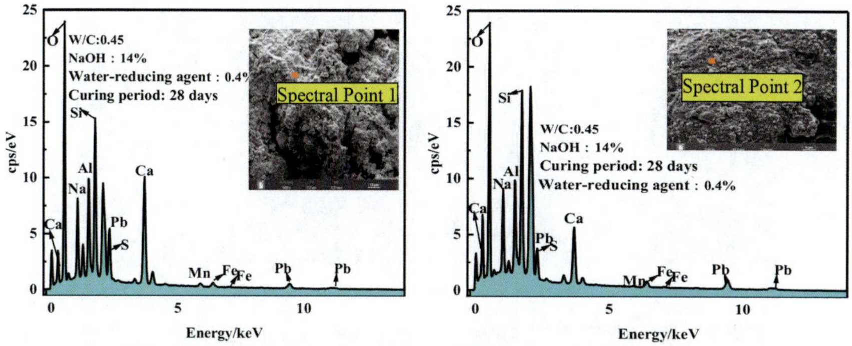

(a) Water-reducing agent ECGM 26-30 (W/C: 0.45)

(b) Water-reducing agent ECGM 36-40 (W/C: 0.45)

Fig. 4.47 EDS results of hardened specimens with different material ratios containing water reducers. (**a**) Water-reducing agent ECGM 26–30 (W/C: 0.45). (**b**) Water-reducing agent ECGM 36–40 (W/C: 0.45)

Table 4.6 Elemental content from Fig. 4.47

Element Types	Spectral Point 1		Spectral Point 2	
	Wt%	σ	Wt%	σ
O	42.5	0.3	43.7	0.2
Si	15.8	0.1	11.8	0.1
Ca	11.1	0.1	18.1	0.1
Na	8.2	0.1	7.0	0.1
Al	7.3	0.1	7.0	0.1
Pb	6.5	0.3	2.9	0.3
S	4.1	0.1	5.5	0.1
Fe	4.1	0.2	2.5	0.1
Mn	0.5	0.1	1.6	0.1
Total volume	100	–	100	–

containing the superplasticizer. Inadequate CSS content can lead to insufficient hydration, thereby reducing the material strength.

Figure 4.48 presents the SEM micrographs and EDS results of specimens with three different mix proportions, both with and without the alkali-free accelerator, at a curing age of 28 days. As shown in Fig. 4.48a, b, the micrographs of the ECGM 41–45 test group with the accelerator reveal the presence of more densely formed geopolymer gels on the surface, along with the development of C-S-H gel. The interweaving growth of C-(A)-S-H gels contributes to a denser structure. Concurrently, some unreacted inert microparticles and plate-like CSS are observed on the specimen surface. The overall structural compactness of this group is lower compared to the ECGM 46–50 and ECGM 51–55 test groups.

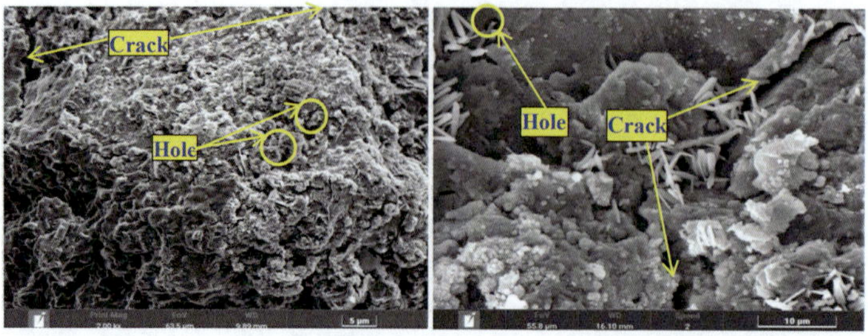

(a) ECGM 41-45 Rapid Setting Agent, ×1000 time (b) Control group, ×5000 magnification

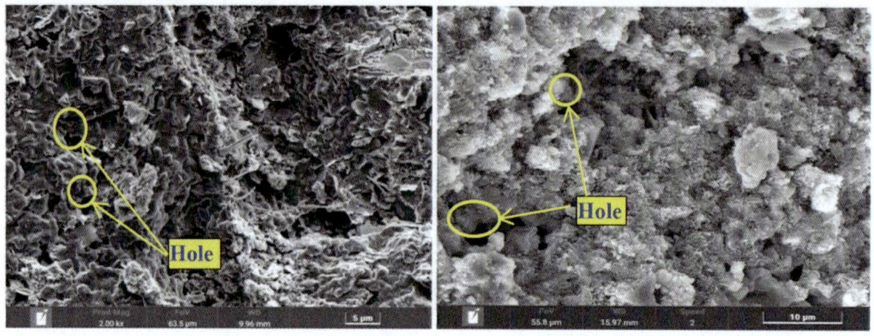

(c) ECGM 46-50 Rapid Setting Agent, ×2000 time (d) Control group, ×5000 magnification

(e) ECGM 51-55 Rapid Setting Agent, ×2000 time (f) Control group, ×5000 magnification

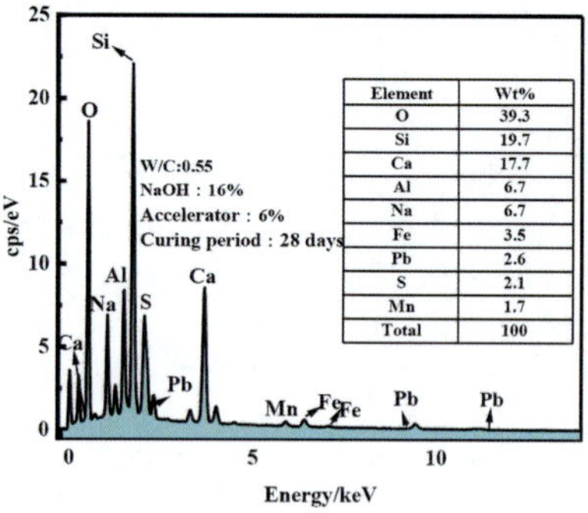

(g) ECGM 51-55 Rapid Setting Agent EDS Results

Fig. 4.48 SEM images and EDS results of cured specimens with different material ratios incorporating accelerators. (**a**) ECGM 41–45 Rapid Setting Agent, ×1000 times. (**b**) Control group, ×5000 magnification. (**c**) ECGM 46–50 Rapid Setting Agent, ×2000 times. (**d**) Control group, ×5000 magnification. (**e**) ECGM 51–55 Rapid Setting Agent, ×2000 times. (**f**) Control group, ×5000 magnification. (**g**) ECGM 51–55 Rapid Setting Agent EDS Results

Furthermore, the micrographs of the ECGM 46–50 and ECGM 51–55 test groups show a greater quantity of formed geopolymer gels, creating a network-like C-S-H gel structure that constitutes the main body of the material strength. The hydration products of C-S-H gel grow in an interlocking manner, with the gels encapsulating unreacted particles such as CSS to form an irregular spatial assembly. Compared to specimens without the accelerator, those containing the accelerator exhibit significantly fewer pores and voids, resulting in a denser microstructure.

According to Fig. 4.48g, the elemental composition of the gel products in the ECGM 51–55 group with the accelerator remains unchanged in terms of element types, but differences in elemental concentrations are observed. This may be due to the influence of the accelerator on the geopolymerization reaction within the matrix, altering the amount of gel products formed in the system and thereby affecting the mechanical strength of the specimens.

In summary, the addition of the polycarboxylate superplasticizer adversely affects the microstructure of the test specimens across all groups. It impairs the compactness of the matrix and reduces the formation of gel products, thereby lowering the mechanical strength. In contrast, the hardened ECGM specimens incorporating the alkali-free accelerator exhibit a denser microstructure and the highest quantity of geopolymer gels formed, resulting in superior strength for the ECGM material. Furthermore, the accelerator enhances the hydration reaction rate of the grout, promotes the rapid formation of geopolymer gel crystals, and effectively shortens the setting time of the slurry.

4.4.3.3 MIP

Figure 4.49 presents the test results of the pore structure characteristics for ECGM specimens from the ECGM 36–40 and ECGM 51–55 test groups, with different admixtures, at a curing age of 28 days. As shown in Fig. 4.49a, the specimen containing the accelerator exhibits the lowest cumulative pore volume curve, followed by the specimen without any admixture, while the specimen with the superplasticizer shows the highest cumulative pore volume curve. This indicates that the addition of the polycarboxylate superplasticizer increases the pore volume of ECGM, whereas a certain amount of the alkali-free accelerator reduces it. The peak cumulative pore volume is highest for the specimen with the superplasticizer, reaching 0.308 cm^3/g, and lowest for the specimen with the accelerator, at 0.277 cm^3/g.

From Fig. 4.49b, it can be observed that the distribution curve of the specimen with the superplasticizer shifts to the right, while that of the specimen with the accelerator shifts toward smaller pore sizes. This suggests that the pore volume increases in specimens with the superplasticizer but decreases in hardened specimens with the accelerator. Compared to the specimen without any admixture, the superplasticizer leads to an increase in porosity, whereas the accelerator results in a decrease. These trends indicate that the superplasticizer has a negative effect on the strength of ECGM, while the accelerator enhances it.

In summary, the addition of the superplasticizer exerts certain negative effects on the mechanical strength and pore structure of the specimens across all test groups.

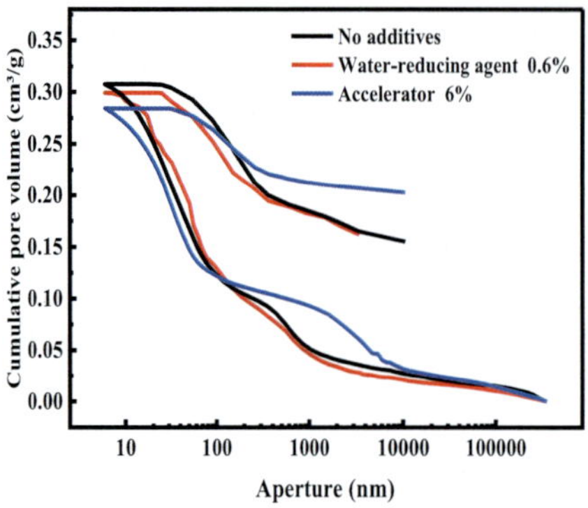

(a) Cumulative pore volume curve

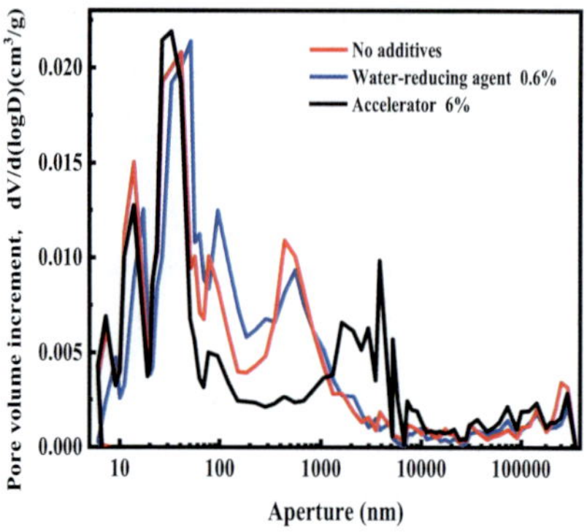

(b) Pore volume increment distribution curve

Fig. 4.49 Effect of different admixtures on the slurry flowability of the ECGM test group. (**a**) Cumulative pore volume curve. (**b**) Pore volume increment distribution curve

However, it effectively improves the fluidity and setting time of the grout. On the other hand, the incorporation of the accelerator reduces the setting time of the grout, enhances the mechanical strength of ECGM, and improves the pore characteristics of the microstructure, although it also adversely affects the flowability of the grout to some extent [45–46].

4.5 Chapter Summary

This chapter investigates the effects of admixtures—specifically a polycarboxylate superplasticizer and an alkali-free accelerator—on the performance of three novel grouting material systems (EFGM, ESGM, and ECGM). These materials were formulated using electrolytic manganese residue (EMR), ground granulated blast-furnace slag (GGBS), fly ash (FA), steel slag (SS), and coal slag (CSS) as raw materials, with NaOH serving as the alkali activator. The influence of admixtures on key properties, including flowability, setting times, uniaxial compressive strength, and microstructural characteristics (via XRD, SEM, and MIP analyses), was systematically evaluated under varying raw material mix proportions. The primary conclusions are as follows:

1. The incorporation of the polycarboxylate superplasticizer significantly enhances the flowability of EFGM, ESGM, and ECGM grouts, with fluidity increasing proportionally with dosage. Conversely, the addition of the alkali-free accelerator reduces grout flowability. Regarding setting times, the superplasticizer prolongs both the initial and final setting times for EFGM and ESGM. For ECGM, setting times first shorten and then lengthen with increasing superplasticizer content. The accelerator extends the setting times of ESGM, while for EFGM and ECGM, setting times initially decrease and then increase with higher accelerator dosage.

2. For the EFGM system, the highest strengths were achieved at the following optimal combinations: 0.6% superplasticizer with 10% EMR, 0.4% superplasticizer with 30% EMR, and 0.8% superplasticizer with 50% EMR. The optimal alkali-free accelerator dosage for EFGM was consistently 4%. In both the ESGM and ECGM systems, compressive strength decreases with increasing superplasticizer content. For the accelerator, compressive strength first increases and then decreases with dosage. The optimal accelerator dosage varies with the EMR content, but an appropriate amount consistently improves compressive strength.
3. The types of hydration products in all three grouting systems are largely unaffected by the addition of either admixture. The primary cementitious products remain C-(A)-S-H and N-A-S-H gels. However, the superplasticizer reduces the alkalinity of the system, decreasing the formation of C-(A)-S-H and N-A-S-H gels and thereby negatively impacting mechanical strength. Porosity and pore volume gradually increase with higher superplasticizer content. In contrast, the alkali-free accelerator effectively reduces pore volume, increases the formation of hydrated cementitious products, and enhances the overall compactness of the system.

References

1. Wang, F., Long, G., Bai, M., et al. (2022). Application of electrolytic manganese residues in cement products through pozzolanic activity motivation and calcination. *Journal of Cleaner Production, 338*, 15.
2. Zou, T., Chen, H., Wang, D., et al. (2023). Experimental study on water washing and high-temperature roasting desulfurization of electrolytic manganese slag. *China Manganese Industry, 41*(01), 48–51.
3. Li, X., Tang, Y., Pan, W., et al. (2023). Study on hydration properties of clinker-silicon manganese slag-limestone composite cementitious material. *Portland Cement Bulletin, 42*(05), 1794–1803.
4. Shi, H., Zhao, Y., & Sha, D. (2000). Mechanism of free calcium oxide on volume expansion of cement paste. *Cement, 04*, 1–3.
5. Cao, H. (2007). *Synthesis and performance study of Polycarboxylic acid superplasticizers*. Anhui University of Science and Technology.
6. Sun, J., & Cao, Y. (2022). Effect of polycarboxylate superplasticizer on the performance of cement mortar. *Shandong Chemical Industry, 51*(22), 16–18.
7. Shaoyun, P. (2021). *Research on Phosphate Base polymer curing agent and its application in curing Lead-contaminated soil*. Southeast University.
8. Jayasree, C., & Gettu, R. (2008). Experimental study of the flow behaviour of superplasticized cement paste. *Materials and Structures, 41*(9), 1581–1593.
9. Zuhua, Z., Mao, Y., Huajun, Z., et al. (2009). Role of water in the synthesis of calcined kaolin-based geopolymer. *Applied Clay Science, 43*(2), 218–223.
10. Song, Y., Zhang, H., & Li, X. (2023). Experimental study on the effect of polycarboxylate superplasticizer on the performance of permeable concrete. *Highway Transportation Technology, 39*(01), 48–53.
11. Zhu, L., Gu, W., Ouyang, Y., et al. (2023). Grouting mechanism and experimental study of goaf considering filtration effect. *PLoS One, 18*(2), 18.
12. Dong, Q., Wang, G., Chen, X., et al. (2021). Recycling of steel slag aggregate in Portland cement concrete: An overview. *Journal of Cleaner Production, 282*, 11.

13. Mo, L., Yang, S., Huang, B., et al. (2020). Preparation, microstructure and property of carbonated artificial steel slag aggregate used in concrete. *Cement & Concrete Composites, 113*, 14.
14. Liu, J., & Guo, R. (2018). Applications of steel slag powder and steel slag aggregate in ultra-high performance concrete. *Advances in Civil Engineering, 2018*, 8.
15. Wang, Q., Yang, J., & Yan, P. (2013). Cementitious properties of super-fine steel slag. *Powder Technology, 245*, 35–39.
16. Su, M., Wang, Z., Zhao, P., et al. (2022). Comparative study on the early strength promotion mechanism of alkaline and non-alkaline liquid accelerators. *Silicate Bulletin, 41*(12), 4172–4179.
17. Salvador R, P., Rambo, D. A. S., Bueno R, M., et al. (2020). Influence of accelerator type and dosage on the durability of wet-mixed sprayed concrete against external sulfate attack. *Construction and Building Materials, 239*, 14.
18. Qu, G., Zheng, M., Zhang, W., et al. (2023). Preparation and acceleration mechanism of a ternary hardening accelerator for high-performance concrete with full aeolian sand. *Construction and Building Materials, 369*, 16.
19. Wang, L., He, X., Shu, C., et al. (2022). Research on the working performance and the corresponding mechanical strength of Polyaluminum sulfate early strength alkali-free liquid accelerator matrix cement. *Materials, 15*(22), 15.
20. Cui, J., He, Z., & Cai, X. (2022). Mechanical properties and key intrinsic factors of sprayed ultra-high performance concrete (SUHPC) with alkali-free accelerator. *Construction and Building Materials, 349*, 14.
21. He, Y., Huang, W., Tang, X., et al. (2023). Experimental study on influencing factors of durability performance of C60 fine aggregate self-compacting concrete. *Journal of Hunan University (Natural Science Edition), 50*(12), 92–101.
22. Huang, B., Cui, S., Wang, Y., et al. (2023). Effect of gypsum dissolution characteristics on the hydration process of anhydrous calcium thioaluminate. *Portland Cement Bulletin, 42*(05), 1804–1813.
23. Rattanasak, U., Pankhet, K., & Chindaprasirt, P. (2011). Effect of chemical admixtures on properties of high-calcium fly ash geopolymer. *International Journal of Minerals, Metallurgy and Materials, 18*(3), 364–369.
24. Sha, F., Li, S., Liu, R., et al. (2018). Experimental study on performance of cement-based grouts admixed with fly ash, bentonite, superplasticizer and water glass. *Construction and Building Materials, 161*, 282–291.
25. Fonseca F, S., Godfrey R, C., & Siggard, K. (2015). Compressive strength of masonry grout containing high amounts of class F fly ash and ground granulated blast furnace slag. *Construction and Building Materials, 94*, 719–727.
26. Saly, F., Guo, L. P., Ma, R., et al. (2018). Properties of steel slag and stainless steel slag as cement replacement materials: A comparative study. *Journal of Wuhan University of Technology-Materials Science Edition, 33*(6), 1444–1451.
27. Xu, S., Ma, Q., Wang, J., et al. (2018). Grouting performance improvement for natural hydraulic lime-based grout via incorporating silica fume and silicon-acrylic latex. *Construction and building Materials, 186*, 652–659.
28. Ren, C., Wang, W., & Li, G. (2017). Preparation of high-performance cementitious materials from industrial solid waste. *Construction and Building Materials, 152*, 39–47.
29. Jiang, N., Du, Y., & Liu, K. (2018). Durability of lightweight alkali-activated ground granulated blast furnace slag (GGBS) stabilized clayey soils subjected to sulfate attack. *Applied Clay Science, 161*, 70–75.
30. Lan, J., Sun, Y., Huang, P., et al. (2020). Using electrolytic manganese residue to prepare novel nanocomposite catalysts for efficient degradation of Azo dyes in Fenton-like processes. *Chemosphere, 252*, 12.
31. Lv, Y., Li, J., Ye, H., et al. (2019). Bioleaching behaviors of silicon and metals in electrolytic manganese residue using silicate bacteria. *Journal of Cleaner Production, 228*, 901–909.
32. Jiang, M., Du, Y., Du, D., et al. (2014). Experimental study on the preparation of silicon-manganese fertilizer using electrolytic manganese slag. *China Manganese Industry, 32*(2), 16–19,24.

33. Xie, C., Xu, L., & Li, L. (2012). Preparation of SiO2-doped manganese-zinc ferrites from electrolytic manganese slag. *Acta In Vitro Ophthalmologica Sinica, 41*(3), 816–820.
34. Liu, W., & Zuo, H. (2021). Effect of MnO and substituting CaO with BaO on the desulfurization ability of blast furnace slag. *Metallurgical and Materials Transactions B, 52*(4), 2275–2282.
35. Nedunuri, A. S. S. S., & Muhammad, S. (2021). Fundamental understanding of the setting behaviour of the alkali activated binders based on ground granulated blast furnace slag and fly ash. *Construction and Building Materials, 291*, 9.
36. Kretzer, M. B. E. C. S. S. (2021). Hybrid geopolymer-cement coating mortar optimized based on metakaolin, fly ash, and granulated blast furnace slag. *Cleaner Engineering and Technology, 4*, 100153.
37. Cao, H. (2022). Effect of water reducer type on the properties of red mud-slag-based polymer grouting materials and its mechanism. *Metal Mining, 2*, 231–236.
38. Wang, M., Wang, J., Gan, J., et al. (2020). Research status on hydration reaction characteristics and setting/solidification mechanisms of the "cement-accelerator-water" system. *Portland Cement Bulletin, 39*(10), 3045–3054.
39. Sheng, Y., Li, G., & Jiang, H. (2020). Effects of water reducer and fly ash on rheological properties of full tailings cemented backfill slurry. *Journal of Chongqing University, 43*(04), 55–63.
40. Gao, W., Wang, Z., Zhao, W., et al. (2015). Preparation and performance study of Fe-Mn-Cu-Co-based infrared radiation materials composite with electrolytic manganese slag. *Functional Materials, 46*(06), 6076–6080.
41. Li, M., Huang, F., Hu, L., et al. (2020). Efficient activation of peroxymonosulfate by a novel catalyst prepared directly from electrolytic manganese slag for degradation of recalcitrant organic pollutes. *Chemical Engineering Journal, 401*, 126081–126085.
42. Ding, Z., Hong, X., Zhu, J., et al. (2018). Study on alkali-activated red mud-slag Geopolymer cement. *Acta Electronica Micrologica Sinica, 37*(02), 145–153.
43. Xu, C., Zhang, Z., Chen, M., et al. (2022). Formation, mechanism of action, and applications of layered bimetallic hydroxides in cement concrete. *Materials Herald, 36*(11), 99–105.
44. Zheng'an, Z. (2011). *Performance study and mechanism exploration of alkali-free liquid accelerators*. Shenyang Jianzhu University.
45. Cai, Y., Liu, X., Kong, X., et al. (2016). Thermal adaptability and coagulation promotion mechanisms of different basic liquid accelerators under high and low temperatures. *Journal of Silicate Science, 44*(11), 1563–1570.
46. Jiezong, G. (2014). *Composition, properties and mechanism research of chlorine-free and alkali-free liquid accelerators*. China Building Materials Science Research Institute.

Part IV
Long-Term Performance and Prediction of Manganese Slag-Based Grouting Materials

Chapter 5
Study on the Durability and Environmental Compatibility of Manganese Slag-Based Grouting Materials

5.1 Introduction

In the field of tunnel and underground engineering, grouting materials are continuously exposed to complex groundwater environments, enduring persistent erosion from groundwater. Particularly, the impact of SO_4^{2-} and Cl^- ions in groundwater on the durability of grouting materials cannot be overlooked [1]. Additionally, manganese slag-based grouting materials pose potential heavy metal contamination risks. As an alkali-activated material, they exhibit relatively high alkali content. When exposed to groundwater, both heavy metal ions and high-concentration alkali components may leach out, contaminating the surrounding environment. Given these concerns, it is necessary to evaluate the leaching of heavy metals and alkali components from manganese slag grouting materials to assess their environmental safety.

This chapter selected analytical-grade Na_2SO_4 and NaCl as solutes for ion erosion testing, preparing solutions with varying ionic concentrations. Strength tests were conducted to investigate the influence patterns of different ion concentrations in erosive solutions on the strength characteristics of novel grouting materials, revealing the intrinsic relationship between ion erosion and material strength. Microstructural testing investigates the influence of varying ionic concentrations on the microstructural properties of the novel grouting material. Toxicity leaching tests examine the leaching patterns of heavy metal ions in manganese slag-based grouting materials, elucidating the solidification mechanism of heavy metal ions within these materials. pH testing characterizes the extent of alkali component leaching. This research examines the leaching characteristics of alkali components in manganese slag grouting materials and the influence of different admixtures on alkali leaching, thereby evaluating the environmental compatibility of manganese slag-based grouting materials.

Z. Wu et al., *Multi-Source Industrial Solid Waste Synergistically Utilized in Manganese Slag-Based Grouting Materials*,
https://doi.org/10.1007/978-981-92-3021-1_5

5.2 Effect of Ion Erosion on Strength Characteristics of Manganese Slag-Based Grouting Materials

5.2.1 Effect of Ion Erosion on Strength Characteristics of EFGM

5.2.1.1 Effect of SO_4^{2-} on the Strength Characteristics of EFGM

The effect of different SO_4^{2-} concentrations on compressive strength is shown in Fig. 5.1. Figure 5.1a presents the compressive strength of grouting material with 10% EMR content, while Fig. 5.1b shows the compressive strength with 30% EMR content. As shown in Fig. 5.1a, when the EMR content is 10%, the strength at curing ages of 14 days and 28 days increases with increasing SO_4^{2-} concentration. Specifically, the 14-day strength exhibits a larger increase with higher SO_4^{2-} concentration, while the 56-day strength decreases as SO_4^{2-} concentration increases. However, at an SO_4^{2-} concentration of 0.8%, the 56-day strength was lower than the 28-day strength, indicating a strength reversal phenomenon. At curing ages of 14 and 28 days, increased SO_4^{2-} concentration accelerated the formation of abundant calcium aluminate hydrates (CAH). These newly formed CAH participated in secondary hydration reactions, further generating monosulfide-type calcium aluminate hydrate gel. This gel material formed a denser microstructure within the material, effectively enhancing interparticle cohesion. Consequently, it significantly improved the material's early-to-mid-stage strength performance, leading to enhanced early-to-mid-stage strength [2, 3].

During the late stages of hydration, the substances within the system that could continue reacting with the calcium aluminate hydrate are gradually depleted. This depletion prevents excess calcium aluminate hydrate from further participating in

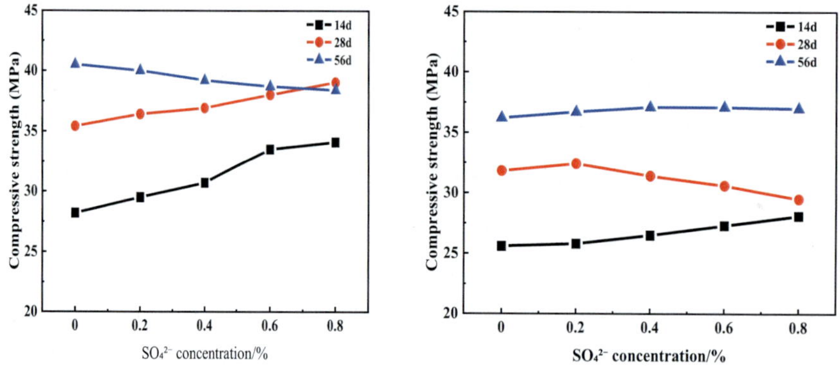

(a) Electrolytic manganese slag content: 100% (b) Electrolytic manganese slag content: 30%

Fig. 5.1 Effect of different SO_4^{2-} concentrations on compressive strength. (**a**) Electrolytic manganese slag content: 100%. (**b**) Electrolytic manganese slag content: 30%

the hydration process. The calcium aluminate hydrate itself possesses water-absorbing and swelling properties. As it continuously absorbs water and expands within the material, this process causes damage to the internal structure. The originally compact structure becomes loose, and the cohesive force between particles decreases. This leads to a reduction in the strength of the specimen.

As shown in Fig. 5.1b, when the EMR content reaches 30%, the significant increase in SO_4^{2-} concentration in the solution has a markedly reduced effect on specimen strength due to the substantial calcium sulfate component present in EMR. At a curing age of 14 days, specimen strength increases with rising SO_4^{2-} concentration. Higher SO_4^{2-} levels further promote relevant chemical reactions, generating more substances that enhance strength and thereby improve overall strength. At a curing age of 28 days, specimen strength first increased then decreased with SO_4^{2-} concentration, peaking at 0.2% SO_4^{2-}. Beyond 0.2%, excess SO_4^{2-} likely interfered with normal hydration processes, inhibiting further strength gain and causing a decline. At a curing age of 56 days, specimen strength increased with SO_4^{2-} concentration before decreasing. The maximum strength was achieved at an SO_4^{2-} concentration of 0.4%. However, excessively high SO_4^{2-} concentrations may disrupt the microstructural equilibrium within the material, leading to strength reduction.

5.2.1.2 Effect of Cl^- on the Strength Characteristics of EFGM

The effect of Cl^- concentration on compressive strength is shown in Fig. 5.2. When the EMR content was 10%, the 14-day strength of the specimens increased with increasing Cl^- concentration. The 28-day strength of the specimens first increased and then decreased with increasing Cl^- concentration, reaching a maximum value at a Cl^- concentration of 0.4%. The 56-day strength of the specimens decreased with increasing Cl^- concentration. Notably, beyond a Cl^- concentration of 0.4%, the specimens exhibited strength shrinkage. The NaCl used in this study is highly soluble in water and rapidly ionizes into Na^+ and Cl^- during erosion. The smaller radius of Na^+ allows it to easily penetrate the specimens, balancing the negative ions generated by hydration reactions within the specimens, while Cl^- reduces the zeta potential in the solution [4], facilitating the decomposition of silica-alumina components and the aggregation of aluminosilicate tetrahedra within the specimen. Thus, a certain concentration of Cl^- enhances the strength of the new grouting material, consistent with the conclusions drawn from the MIP test. As Cl^- concentration increases, both Na^+ and Cl^- concentrations rise within the grouting material. Excessively high Na^+ levels can bind with oxygen atoms on the surface of silicoaluminates in the grout, triggering a "passivation" phenomenon [4]. This subsequently hinders the decomposition of silicoaluminate phases in the grouting material under alkaline conditions, thereby impeding the hydration reaction process. Furthermore, excessive Cl^- precipitates as crystals within the grouting material. This not only hinders ion migration during hydration, limiting reaction occurrence, but also weakens the

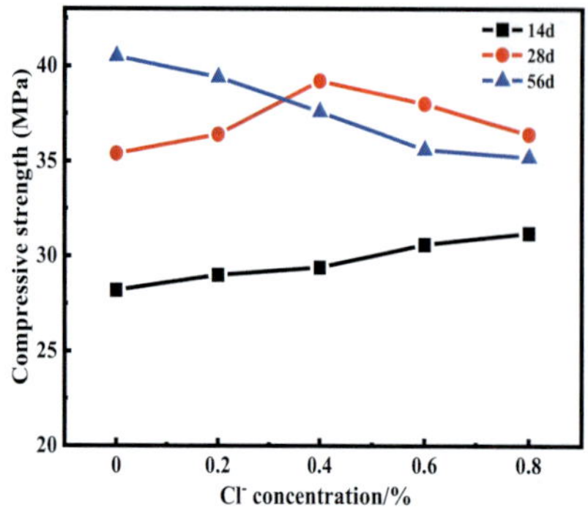

(a) Electrolytic manganese slag content: 10%

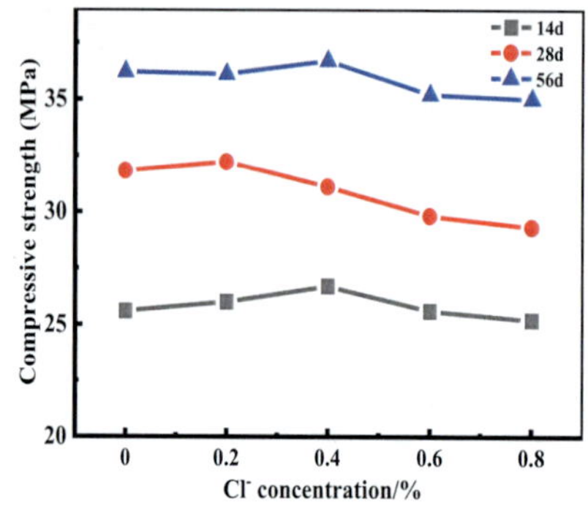

(b) Electrolytic manganese slag content: 30%

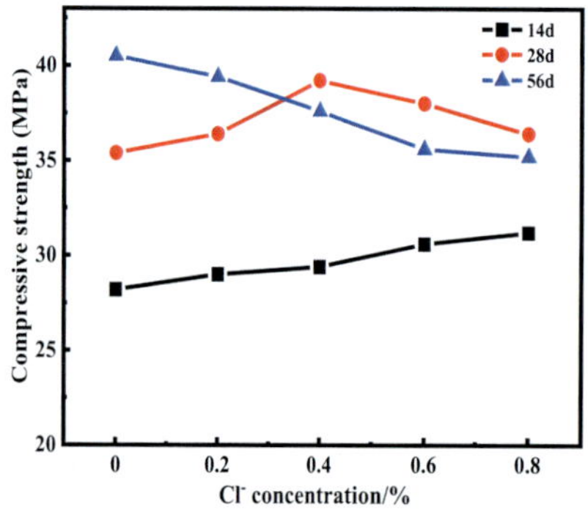

(a) Electrolytic manganese slag content: 10%

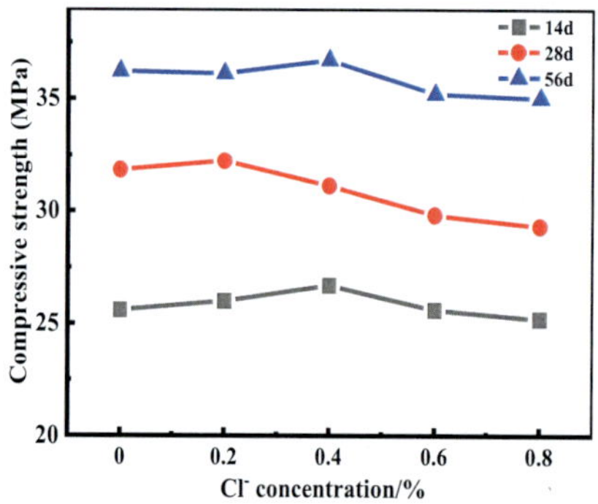

(b) Electrolytic manganese slag content: 30%

Fig. 5.2 Effect of different Cl^- concentrations on compressive strength. (**a**) Electrolytic manganese slag content: 10%. (**b**) Electrolytic manganese slag content: 30%

binding capacity of hydration products. Ultimately, this increases pore volume and radius in the new grouting material, leading to reduced strength.

When the EMR content reaches 30%, the weakening effect of Cl^- on specimen strength is significantly reduced. When the Cl^- concentration does not exceed 0.4%, it promotes the compressive strength of the specimens. At a manganese slag content of 30%, the abundant calcium sulfate in EMR induces sulfate activation of GGBS. GGBS activated by sulfates exhibits superior resistance to Cl^- erosion, preventing Cl^- from negatively impacting material strength within a certain concentration range and instead enhancing strength.

5.2.2 *Effect of Ion Erosion on the Strength Characteristics of ESGM*

5.2.2.1 Effect of SO_4^{2-} on the Strength Characteristics of ESGM

The comparison of specimen strength under SO_4^{2-} solution erosion and standard curing conditions is shown in Fig. 5.3. For specimens with 10% EMR content, $f_{(erod\text{-}compressive)}(n)$ was slightly greater than $f_{(standard\text{-}compressive)}$ (n) from 3 to 14 days. For specimens with 20% EMR content, $f_{(corr)\text{-}compressive}$ (n) decreased and remained below $f_{(spec)\text{-}compressive}$ (n) from 0 to 3 days. From 3 to 14 days, $f_{(corr)\text{-}compressive}$ (n) began to increase, reaching parity with $f_{(spec)\text{-}compressive}$ (n) at 14 days. From 14 to 56 days, $f_{(corr)\text{-}compressive}$ (n) continued to increase and exceeded $f_{(spec)\text{-}compressive}$ (n). For specimens with 30% EMR content: From 0 to 3 days, $f_{(intrusion)\text{-}compressive\ strength}$ (n) decreased and remained below $f_{(standard)\text{-}compressive\ strength}$ (n); From 3 to 14 days, $f(_{intrusion)\text{-}compressive\ strength}$ (n) began to increase but remained below $f_{(standard)\text{-}compressive\ strength}$ (n); – From 14 to 28 days, $f_{(intrusion)\text{-}compressive\ strength}$ (n) continued to increase and exceeded $f(_{standard)\text{-}compressive\ strength}$ (n) at 28 days; From 28 to 56 days, $f_{int\text{-}compressive}$ (n) continued to increase and exceeded $f_{spec\text{-}compressive}$ (n). Specimens with 40% EMR content exhibited strength changes similar to those with 20% EMR content, with $f_{int\text{-}compressive}$ (n) gradually exceeding $f_{spec\text{-}compressive}$ (n) between 3 and 14 days. For specimens with 50% EMR content, $f_{(corrosion)\text{-}compressive\ strength}$ (n) decreased from 0 to 3 days, showed a slow increase from 3 to 56 days, and ultimately remained below $f_{(spec)\text{-}compressive\ strength}$ (n). This indicates that when EMR content ≤40%, SO_4^{2-} solution corrosion promotes compressive strength.

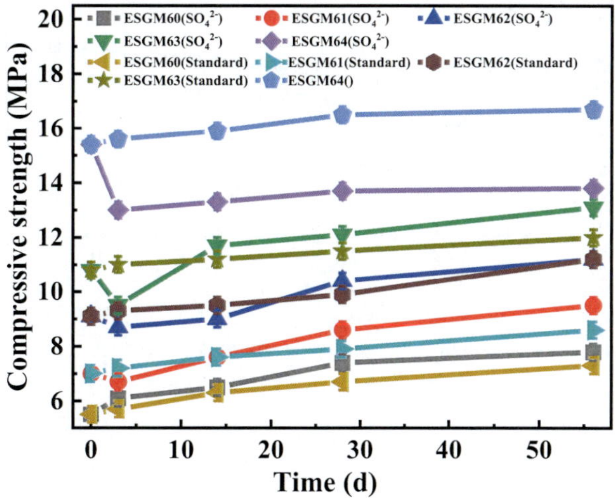

Fig. 5.3 Comparison of specimen strengths under standard curing and Na_2SO_4 solution erosion

5.2.2.2 Effect of Cl^- on the Strength Characteristics of ESGM

Figure 5.4 compares the strength of specimens subjected to Cl^- solution erosion with those under standard curing conditions. At EMR dosages of 10%, 20%, and 30%, the compressive strength ($f_{erod\text{-}compressive}$ (n)) of specimens increased with erosion age. At 56 days, f($_{erod\text{-}compressive}$ (n)) was not significantly different from f($_{spec\text{-}compressive}$ (n)), indicating strong resistance to chloride ion erosion. For an EMR content of 40%, $f_{(corr)\text{-}compressive\ strength}$ (n) exceeded $f_{(spec)\text{-}compressive\ strength}$ (n) from 0 to 3 days, decreased between 3 and 14 days, fell below $f_{(spec)\text{-}compressive\ strength}$ (n) at 14 days, and gradually increased from 14 to 56 days, equaling $f_{(spec)\text{-}compressive\ strength}$ (n) at 56 days. This demonstrates good resistance to chloride ion corrosion. For an EMR content of 50%, $f_{corrosion\text{-}compressive\ strength}$ (n) remained below $f_{standard\text{-}compressive\ strength}$ (n) throughout the entire corrosion aging period. It decreased from 0 to 3 days, increased from 3 to 14 days, dropped significantly from 14 to 28 days, and showed slight growth from 28 to 56 days, indicating weaker resistance to Cl^- corrosion. This indicates that when the EMR content is ≤40%, the corrosive effect of Cl^- on compressive strength is minor; however, when the EMR content is ≥40%, the corrosive effect of Cl^- on compressive strength is significant.

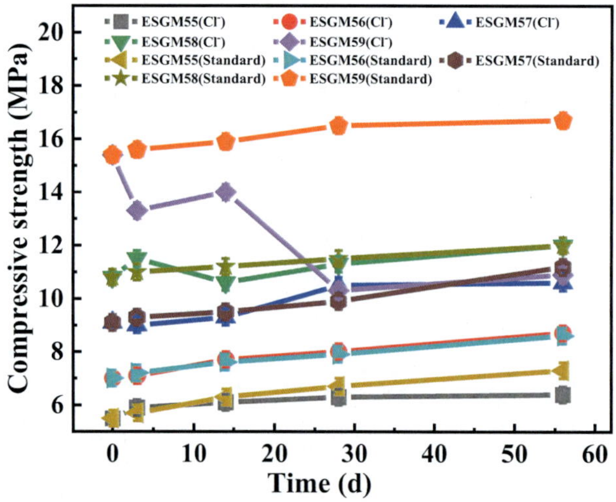

Fig. 5.4 Comparison of strength between specimens under standard curing and NaCl solution erosion

5.2.3 *Effect of Ion Erosion on ECGM Intensity Characteristics*

5.2.3.1 Effect of SO_4^{2-} on ECGM Strength Characteristics

Figure 5.5 shows the strength comparison between specimens exposed to SO_4^{2-} solution and those under standard curing conditions. Specimens immersed in SO_4^{2-} solution exhibited reduced strength, with the difference from standard-cured specimens increasing over curing age. This indicates a negative impact of SO_4^{2-} solution on specimen mechanical strength. SO_4^{2-} ions affect the mechanical properties of hardened specimens. Within the 28-day age range, specimens under both curing conditions exhibited slight strength increases. This may be attributed to two factors: incomplete hydration reactions within the specimens, leaving residual unreacted hydration products; and the salt-stimulating effect of trace SO_4^{2-} ions on solid waste raw materials such as EMR, CSS, and GGBS, which may promote hydration reactions and enhance grouting material strength. Additionally, trace SO_4^{2-} ions react with C-S-A-H gel to form calcium aluminate hydrate (CAH) [5]. A small amount of calcium aluminate silicate hydrate enhances strength and reduces cracking in the system. However, excessive SO_4^{2-} ions negatively impact strength development. Excess SO_4^{2-} ions react with hydration products like C-S-A-H gel, forming gypsum and calcium aluminate silicate hydrate. This generates cracks within the matrix, compromising the specimen's density and consequently reducing strength.

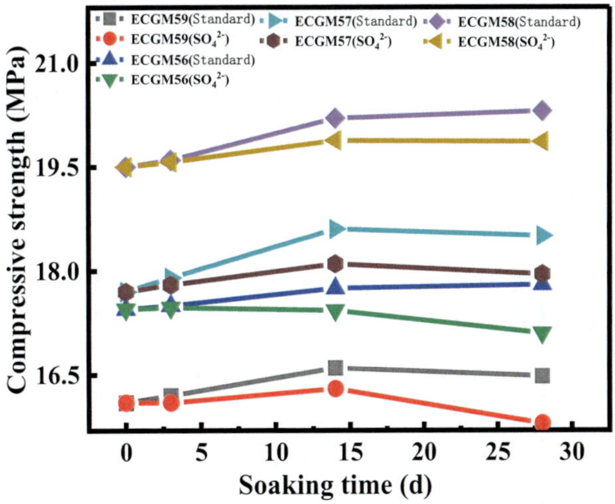

Fig. 5.5 Comparison of strength under standard curing and SO_4^{2-} erosion conditions

5.2.3.2 Effect of Cl^- on ECGM Intensity Characteristics

Table 5.1 shows the variation patterns of strength erosion coefficients for different material mix ratios under SO_4^{2-} solution erosion conditions. As immersion time increased, the strength erosion coefficients of all specimens decreased to varying degrees. The ECGM62 group exhibited the best resistance to SO_4^{2-} solution erosion, followed by the ECGM61 group. The most pronounced erosion occurred in the ECGM64 group. An increase in CSS dosage led to a strength erosion coefficient that first increased and then decreased, with the specimens' resistance to SO_4^{2-} ion erosion following a similar pattern. Furthermore, comparison with groundwater solution test results indicates that material strength decreases under SO_4^{2-} ion erosion conditions.

5.2.3.3 Effect of Cl^- on ECGM Strength Characteristics

Figure 5.6 compares the strength of specimens under standard curing conditions with those subjected to Cl^- solution erosion. Specimens cured under standard conditions consistently exhibit higher strength than those eroded in Cl^- solutions. As immersion time increases, the strength difference between the two groups widens,

Table 5.1 Strength erosion coefficients of different slurry material mix proportions (Na_2SO_4 Solution)

	Slurry Type			
	ECGM60	ECGM61	ECGM62	EGCM63
Soaking time	Stress-resistant	Stress-resistant	Stress-resistant	Stress-resistant
3d	0.992	0.994	0.995	0.990
14d	0.975	0.979	0.981	0.964
28d	0.953	0.964	0.973	0.948

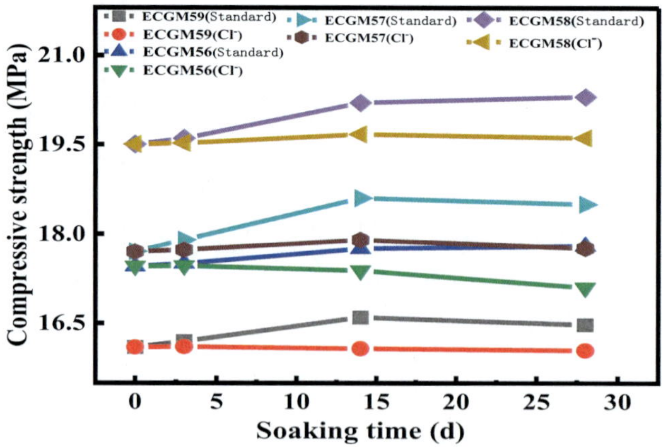

Fig. 5.6 Comparison of strength under standard curing conditions and Cl^- erosion conditions

Table 5.2 Strength erosion coefficients for different grout mix proportions (NaCl solution)

	Slurry Type			
	ECGM56	ECGM57	ECGM58	ECGM59
Soaking time	Stress-resistant	Stress-resistant	Stress-resistant	Stress-resistant
3d	0.981	0.984	0.987	0.974
14d	0.964	0.967	0.974	0.951
28d	0.939	0.945	0.957	0.913

indicating greater erosion effects. Comparing Table 5.5 reveals that at equivalent ages, specimens exposed to Cl^-corrosion solutions exhibit the lowest strength, indicating Cl^- exerts a greater impact on flexural strength than compressive strength. As a highly soluble salt, excessive Na^+ concentration in NaCl solutions induces a "passivation" effect with surface O atoms of silicoaluminates [190], inhibiting raw material dissolution. Additionally, Cl^- can saturate the solution, leaving residual ions within the specimen. This reduces hydration reactions and negatively impacts the bonding between gel products within the system. Increased porosity adversely affects the microstructure, ultimately compromising specimen strength.

Table 5.2 shows the variation patterns of erosion coefficients for different material mixes under chloride-containing solution erosion conditions. As shown in the table, the strength erosion coefficients of all material mixes decrease with increasing immersion time. Among them, the ECGM58 group exhibits the highest compressive and flexural strength erosion coefficients, indicating the best resistance to chloride ions. The ECGM57 group follows, while the ECGM59 group demonstrates the poorest chloride resistance. This demonstrates that incorporating an appropriate amount of CSS effectively enhances the durability of the new grouting slurry against Cl^- corrosion. Furthermore, comparing the corrosion results of groundwater solutions and SO_4^{2-} solutions on the specimens, along with the magnitude of the strength-corrosion coefficients, reveals that Cl^- solutions exert a greater corrosive effect on all specimen groups. Therefore, the alkali-activated new grouting slurry exhibits the poorest resistance to Cl^- corrosion.

5.3 Effect of Ion Erosion on the Microstructural Characteristics of Manganese Slag-Based Grouting Materials

5.3.1 Effect of Ion Erosion on the Microstructural Characteristics of EFGM

5.3.1.1 XRD

SO_4^{2-} Concentration

With a water-cement ratio of 0.4, electrolytic manganese slag content of 30%, and a curing age of 28 days, the XRD patterns of specimens with different SO_4^{2-} concentrations

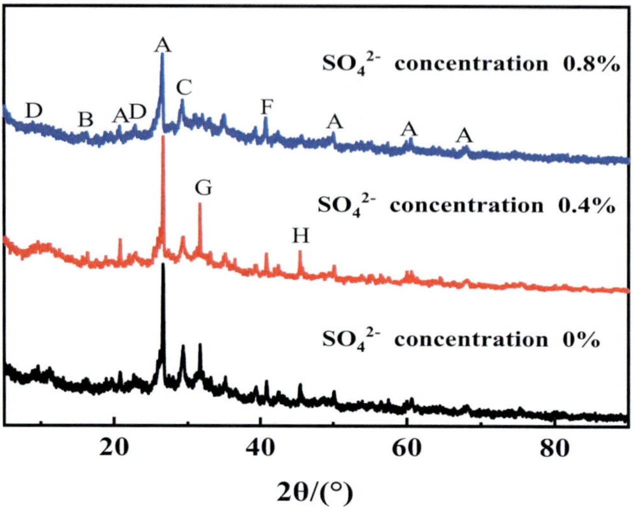

Fig. 5.7 XRD Pattern Analysis of Samples at Different SO_4^{2-} Ion Concentrations (A: Quartz-79-1906, B: Mullite-79-1453, C: Calcium Carbonate-72-1937, D: Calcium Aluminate-41-1451, E: Calcium Hydroxide-87-0673, F: Calcium silicate-23-1042, G: Sodium sulfate-99-0103, H: Sodium chloride-75-0306)

are shown in Figs. 5.7. The primary hydration products are C-S-(A)-H gel and N-(C)-S-A-H gel, along with unreacted quartz, mullite, and hydration products such as calcium carbonate, calcium aluminate hydrate, slaked lime, calcium silicate, sodium sulfate, and sodium chloride (NaCl, PDF#75–0306). As sulfate ion concentration increases, the diffraction peaks of C-S-(A)-H gel and N-(C)-S-A-H gel within the 5–10° and 20–30° ranges intensify, while the characteristic peaks of quartz and mullite weaken. This indicates that a certain sulfate ion concentration is beneficial for the formation of geopolymer gel, promoting the hydration reaction process. At sulfate concentrations of 0% and 0.4%, characteristic peaks of sodium sulfate appeared, indicating incomplete sulfate consumption at these ion concentrations. At 0.8% concentration, characteristic peaks of calcium aluminate, calcium carbonate, and calcium silicate intensified.

With a water-cement ratio of 0.4 and electrolytic manganese slag content of 30%, with a SO_4^{2-} concentration of 0.4%. The XRD patterns of specimens at different curing ages are shown in Figs. 5.8. At a curing age of 14 days, characteristic peaks of calcium aluminate hydrate (calcium aluminate hydrate) appeared. Under sulfate-attacking conditions, a large amount of calcium aluminate hydrate formed during the early hydration stage. Since the secondary hydration reaction of calcium aluminate hydrate was not fully consumed, the characteristic peaks gradually disappeared as the curing age increased. At 28-day curing age, characteristic peaks of sodium sulfate appear while those of mullite weaken, indicating continuous consumption of mullite through hydration reactions. At 56-day curing age, characteristic peaks of sodium chloride appear. This may result from excess Na^+ in the late-stage

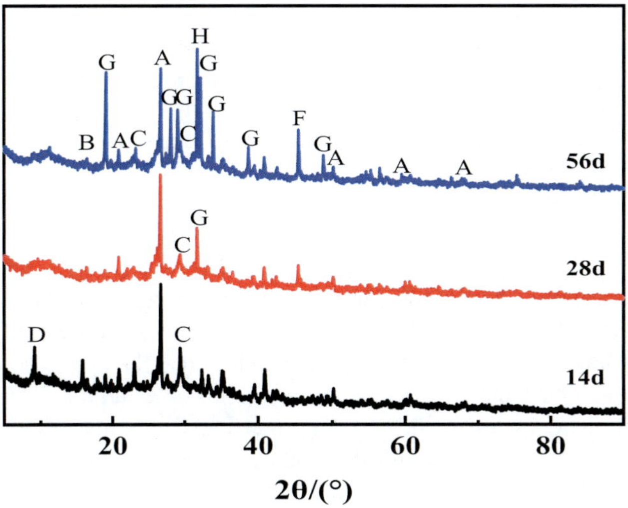

Fig. 5.8 XRD spectrum analysis of specimens at different curing ages (A: Quartz-79-1906, B: Mullite-79-1453, C: Calcium Carbonate-72-1937, D: Calcium Aluminate-41-1451, E: Calcium Hydroxide-87-0673, F: Calcium Silicate-23-1042, G: Sodium Sulfate-99-0103)1) XRD

Fig. 5.9 XRD pattern analysis of samples with different electrolytic manganese slag contents (A: Quartz-79-1906, B: Mullite-79-1453, C: Calcium Carbonate-72-1937, D: Calcium Aluminate-41-1451, E: Calcium Hydroxide-87-0673, F: Calcium Silicate-23-1042, G: Sodium Sulfate-99-0103)

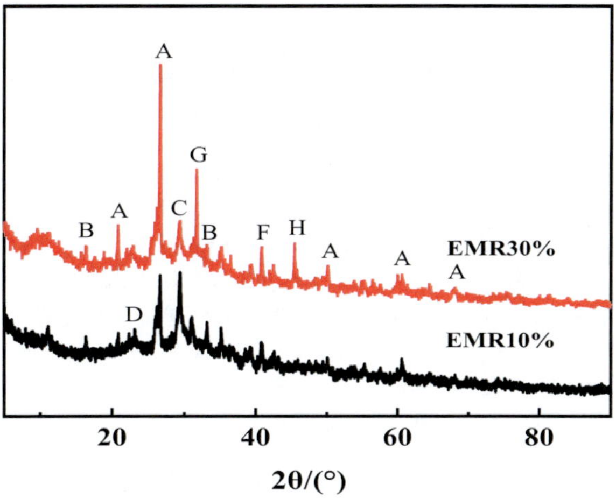

hydration system combining with Cl^- from raw materials to form NaCl crystals. Characteristic peaks of sodium sulfate and sodium silicate progressively intensify.

Figure 5.9 shows the XRD patterns of specimens with different manganese slag contents at a water-cement ratio of 0.4 and a curing age of 28 days, where the SO_4^{2-} concentration was 0.4%. As the manganese slag content increases, the quartz characteristic peak becomes stronger, while the characteristic peak for calcium carbonate weakens. This may occur because increased electrolytic manganese slag content raises the amount of unreacted quartz in the system while lowering Ca^{2+}

concentration. Alternatively, higher slag content may negatively impact the hydration process, reducing quartz consumption. At a manganese slag content of 30%, characteristic peaks for sodium chloride and sodium sulfate appear, while the characteristic peak for sodium silicate intensifies. This may be due to the presence of excess sodium ions, sulfate ions, and silicon ions in the system.

Cl^- Concentration

With a water-cement ratio of 0.4, electrolytic manganese slag content of 30%, and a curing age of 28 days, the XRD patterns of specimens with different Cl^- concentrations are shown in Fig. 5.10. The figure indicates that the hydration products of specimens under different Cl^- concentrations exhibit little variation, demonstrating that Cl^- plays a minor role in the hydration reaction. The primary hydration products are C-S-(A)-H gel and N-(C)-S-A-H gel, alongside unreacted quartz and mullite, as well as hydration products such as calcium carbonate, calcium aluminate hydrate, slaked lime, calcium silicate, sodium sulfate, and sodium chloride. As Cl^- concentration increases, the diffraction peaks of C-S-(A)-H gel and N-(C)-S-A-H gel at 5.10 °C and 20–30 °C become more pronounced, indicating that Cl^- positively influences gel formation, consistent with strength characteristics. As ion concentration increases, the characteristic peak of sodium chloride in the system intensifies, while the characteristic peak of quartz diminishes.

The water-cement ratio is 0.4, the electrolytic manganese slag content is 30%, and the Cl^- concentration is 0.4%. The XRD patterns of specimens at different curing ages are shown in Fig. 5.11. The types of hydration products showed little variation across different curing ages. As curing age increased, the diffraction peaks of C-S-(A)-H and N-(C)-S-A-H gels in the 5–10° and 20–30° ranges progressively strengthened. This indicates a continuous increase in the formation of geopolymer gels, leading to enhanced strength and a gradually denser structure. These findings are consistent with the results obtained from strength and SEM tests. At curing ages

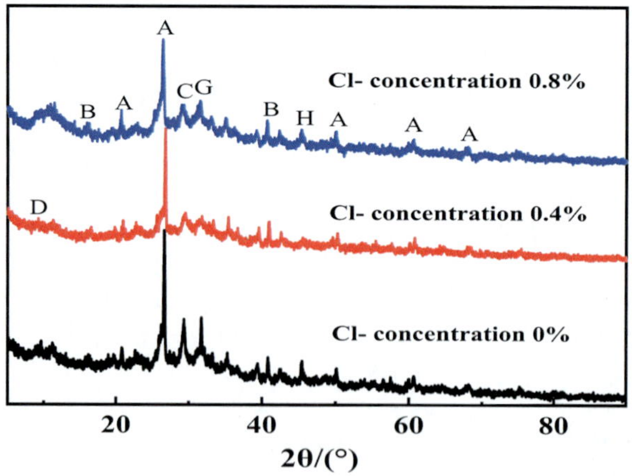

Fig. 5.10 XRD pattern analysis of samples at different Cl^- ion concentrations (A: Quartz-79-1906, B: Mullite-79-1453, C: Calcium Carbonate-72-1937, D: Calcium Aluminate-41-1451, E: Calcium Hydroxide-87-0673, F: Calcium silicate-23-1042, G: Sodium sulfate-99-0103, H: Sodium chloride-75-0306)

Fig. 5.11 XRD spectrum analysis of specimens at different curing ages (A: Quartz-79-1906, B: Mullite-79-1453, C: Calcium Carbonate-72-1937, D: Calcium Aluminate Hydrate-41-1451, E: Calcium Hydroxide-87-0673, F: Calcium Silicate-23-1042, G: Sodium Sulfate-99-0103, H: Sodium Chloride-75-0306)

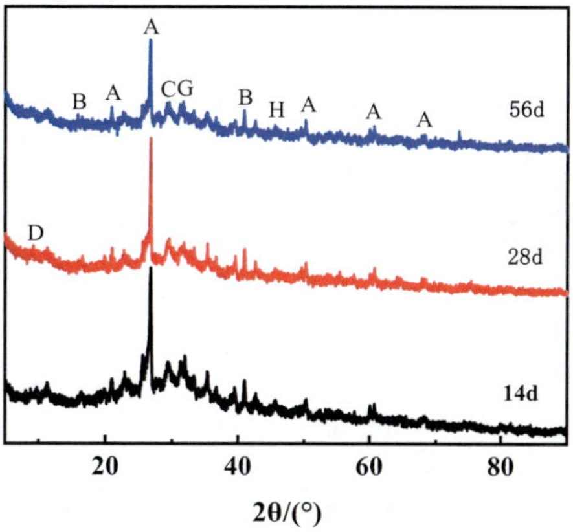

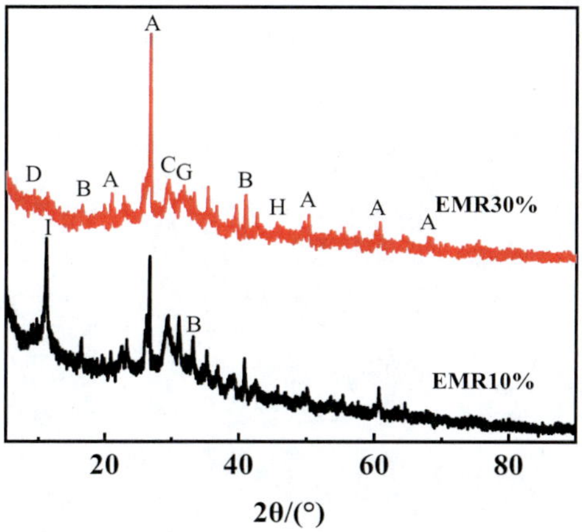

Fig. 5.12 XRD pattern analysis of samples with different EMR content (A: Quartz-79-1906, B: Mullite-79-1453, C: Calcium Carbonate-72-1937, D: Calcium Aluminate-41-1451, E: Calcium Hydroxide-87-0673, F: Calcium silicate-23-1042, G: Sodium sulfate-99-0103, H: Sodium chloride-75-0306, I: Tetramethylammonium chloride-43-1757)

of 14d and 28d, characteristic peaks of calcium aluminate hydrate (calcium aluminate hydrate) are observed. As the curing age increases, the characteristic peaks of calcium aluminate hydrate disappear. This is attributed to chloride ion erosion: at 14 and 28 days, the calcium aluminate hydrate formed during hydration was not fully consumed, but by 28 days, it was completely consumed through secondary hydration reactions. It can be observed that the characteristic peaks of quartz and mullite continuously decrease with increasing curing age, consistent with the progression of the hydration reaction.

The water-cement ratio was 0.4, the curing age was 28 days, and the Cl^- concentration was 0.4%. The XRD patterns of specimens with different manganese slag contents are shown in Fig. 5.12. As seen in the figure, when the electrolytic manganese slag content was 10%, a new hydration product, tetramethylammonium chloride ($C_4H_{12}CdCl_3N$, PDF#43–1757) appears at 11.18°. This may be attributed to Cl^- participating in the hydration reaction under ionic erosion when the manganese slag content is low. As the electrolytic manganese slag content increases, the diffraction peaks of the C-S-(A)-H gel and N-(C)-S-A-H gel at 5–10° and 20–30 °C weaken, while the characteristic peak of quartz intensifies.

5.3.1.2 SEM-EDS

SO_4^{2-} Concentration

The water-cement ratio was 0.4, the curing age was 28 days, and the electrolytic manganese slag content was 30%. The microstructures of specimens with different SO_4^{2-} concentrations are shown in Fig. 5.13. Under ionic erosion, the specimen surfaces were densely covered by hydration products, which also explains the low

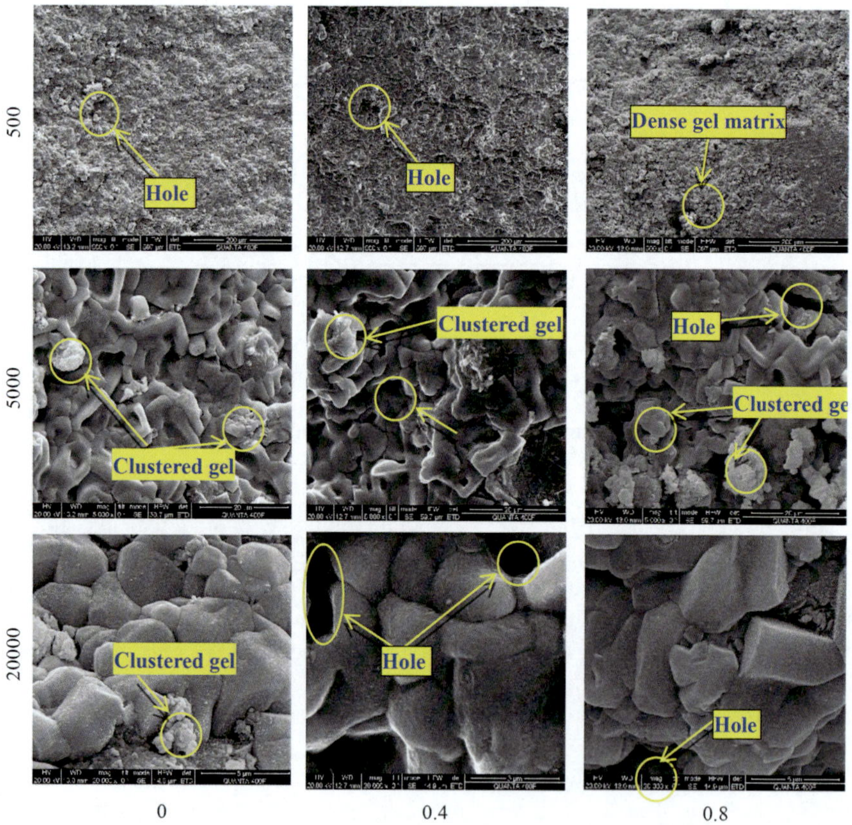

Fig. 5.13 Microstructural morphology of specimens at different SO_4^{2-} concentrations

number of large pores (greater than 200 nm) observed in the specimens. The specimen surface is tightly covered by the generated gel, which reduces the specimen permeability. This is consistent with the results obtained from the MIP test, where the specimen permeability continuously decreases as the ion concentration increases.

The water-cement ratio is 0.4, with an electrolytic manganese slag content of 30%, with an SO_4^{2-} concentration of 0.4%. The microstructures of specimens at different curing ages are shown in Fig. 5.14. The figure indicates that the microstructures of hydration products at different ages exhibit little variation. Micropores and cracks are visible during the early hydration stage. As the curing age increases, the pores in the specimens are gradually filled by newly formed gel, resulting in a denser microstructure.

The water-cement ratio was 0.4, the curing age was 28 days, and the SO_4^{2-} concentration was 0.4%. The microstructures of specimens with different manganese slag contents are shown in Fig. 5.15. When the manganese slag content was 10%, distinct microcracks were visible at magnifications of 5000x and 20,000x. At 30% manganese slag content, microcracks disappeared but micro-voids became visible, and the structure loosened—consistent with MIP specimen results.

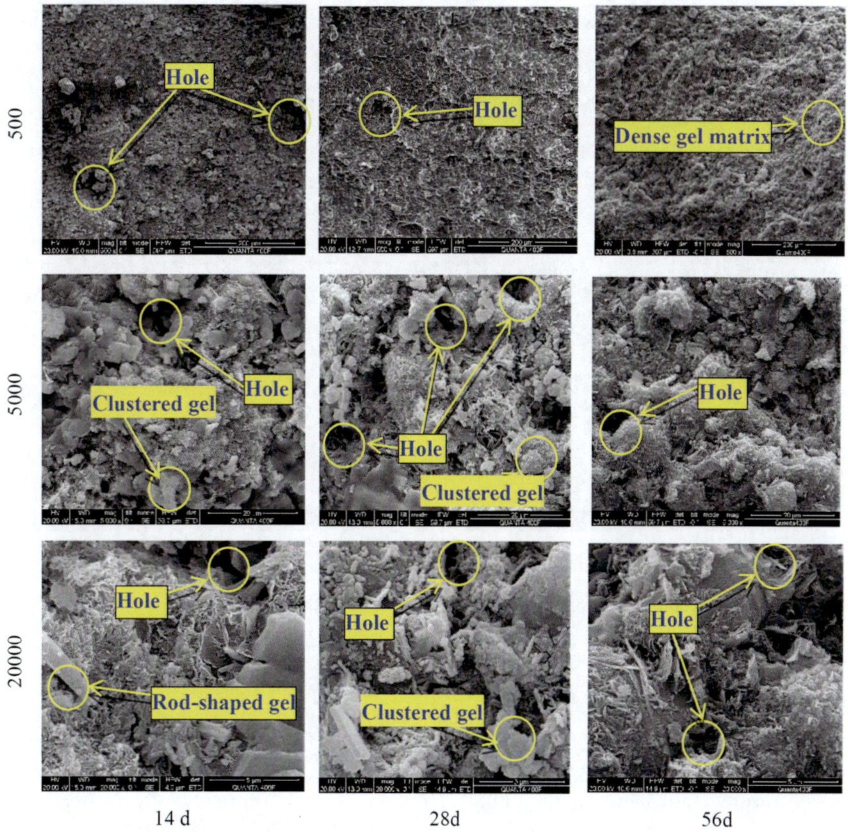

Fig. 5.14 Microstructural morphology of specimens at different curing ages

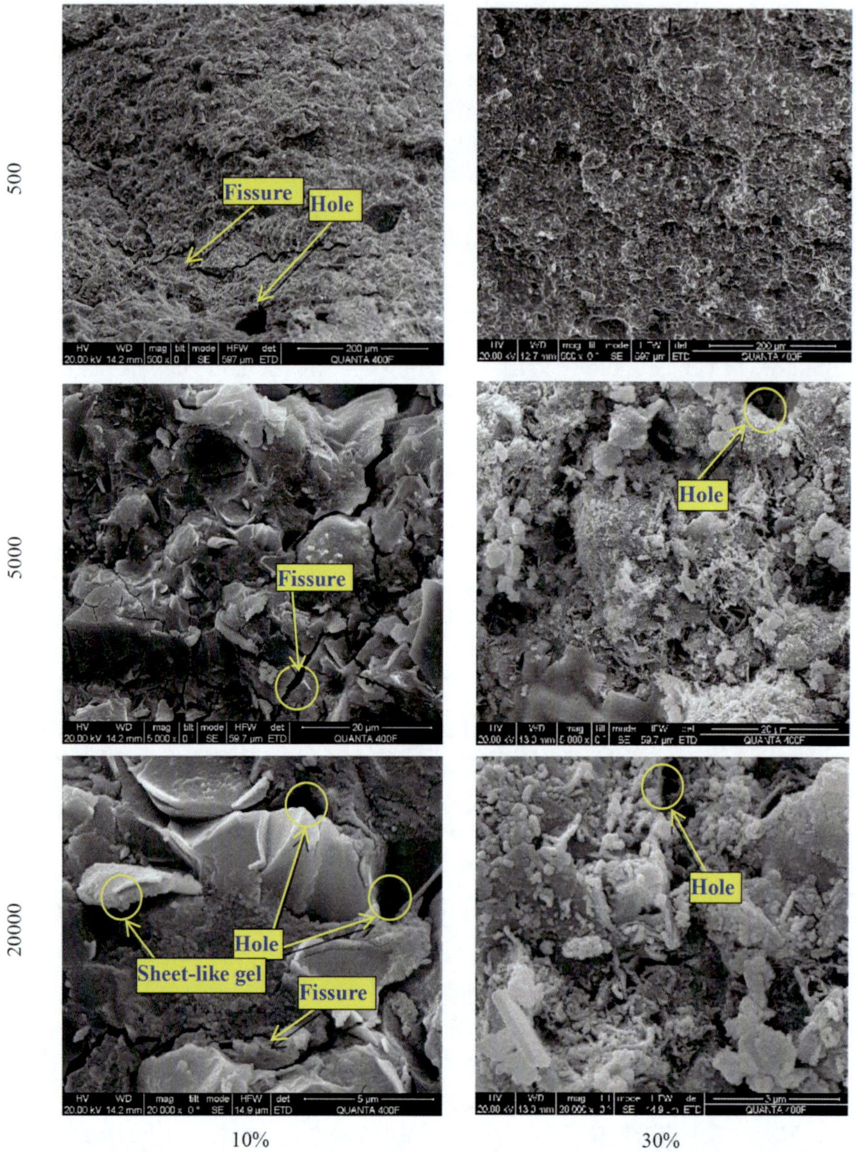

Fig. 5.15 Microstructural morphology of specimens with different manganese slag contents

Cl^- Concentration

With a water-cement ratio of 0.4, a curing age of 28 days, and an electrolytic manganese slag content of 30%, the microstructures of specimens with different Cl^- concentrations are shown in Fig. 5.16. As the Cl^- concentration increases, micro-voids appear between the dense gel phases at 20,000× magnification, leading to an increase in the specimens' porosity.

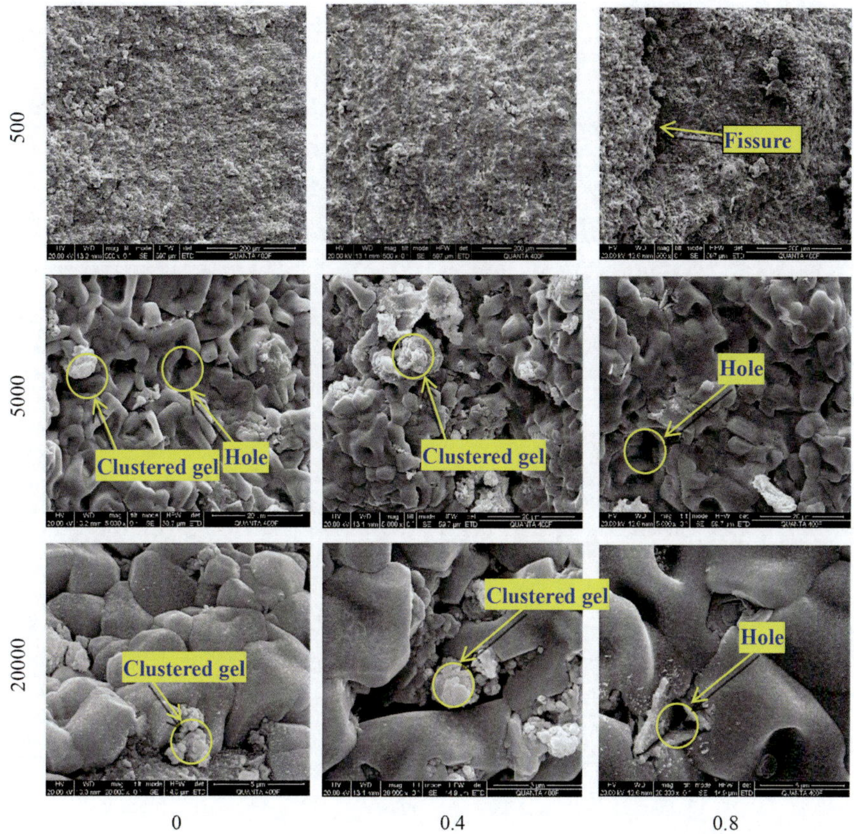

Fig. 5.16 Microstructural morphology of specimens at different Cl^- concentrations

The water-cement ratio was 0.4, the electrolytic manganese slag content was 30%, and the Cl^- concentration was 0.4%. The microstructures of specimens at different curing ages are shown in Fig. 5.17. As the curing age increased, the microvoids between the hydration products in the specimens were gradually filled by newly formed hydration products at magnifications of 5000× and 20,000×. resulting in progressively denser microstructures. This trend aligns with the strength test results.

The water-cement ratio was 0.4, the curing age was 28 days, and the Cl^- concentration was 0.4%. The microstructures of specimens with different manganese slag contents are shown in Fig. 5.18. When the manganese slag content was 10%, a dense microstructure with microcracks was observed. At 30% manganese slag content, pores appeared within the flocculent gel, and the specimen's microstructure gradually loosened. As the electrolytic manganese slag content increased, the specimen's microstructure progressively became more loose, consistent with the results of the specimen's strength tests and MIP tests.

Fig. 5.17 Microstructural morphology of specimens at different curing ages

SO_4^{2-} Concentration

Parameters for specimens with varying SO_4^{2-} concentrations are shown in Table 5.3. When the electrolytic manganese slag content is 30%, the specimen porosity, average pore size, and apparent density first increase and then decrease with increasing SO_4^{2-} concentration; permeability decreases.

The effect of different SO_4^{2-} concentrations on the pore structure of specimens is shown in Fig. 5.19. As the ion concentration increases, the cumulative pore volume first rises and then decreases. Within the 5–100 nm range, the incremental pore volume also first increases and then decreases. The variation in porosity is primarily attributed to changes in the incremental pore volume within the 5–100 nm range.

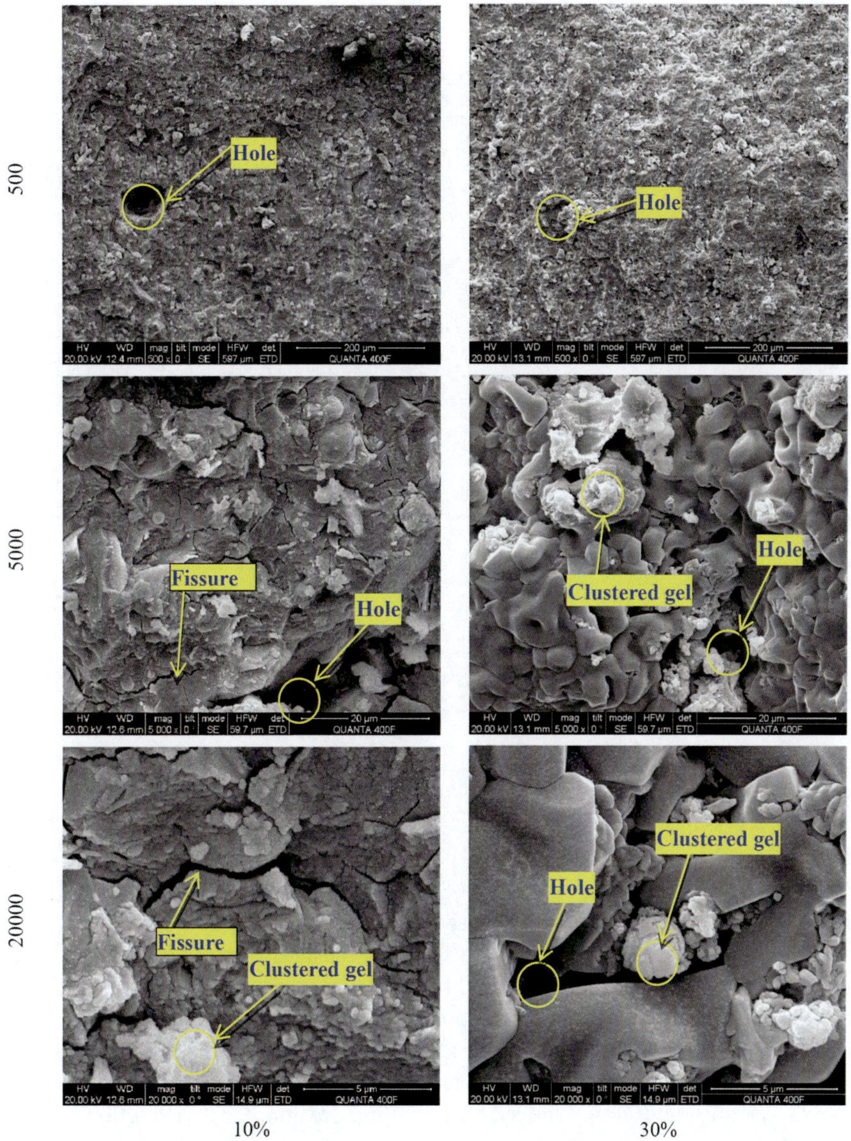

Fig. 5.18 Microstructural morphology of specimens with different manganese slag contents

5.3.2 Effect of Ion Erosion on the Microstructural Characteristics of ESGM

5.3.2.1 XRD

The XRD patterns of ESGM samples under erosion are shown in Fig. 5.20. The primary mineral phases are identified as: A: SiO_2 (PDF#85–0335), B: vanadite

Table 5.3 Parameters of samples at different SO_4^{2-} concentrations

Manganese slag content/%	Water-cement ratio	SO_4^{2-}/%	Average pore diameter/nm	Apparent density/(g/mL)	Porosity/%	Permeability/mD
30	0.4	0	33.361	2.2748	36.5983	406.2041
30	0.4	0.4	33.823	2.359	38.6624	90.7718
30	0.4	0.8	32.221	2.2814	37.1386	64.3896

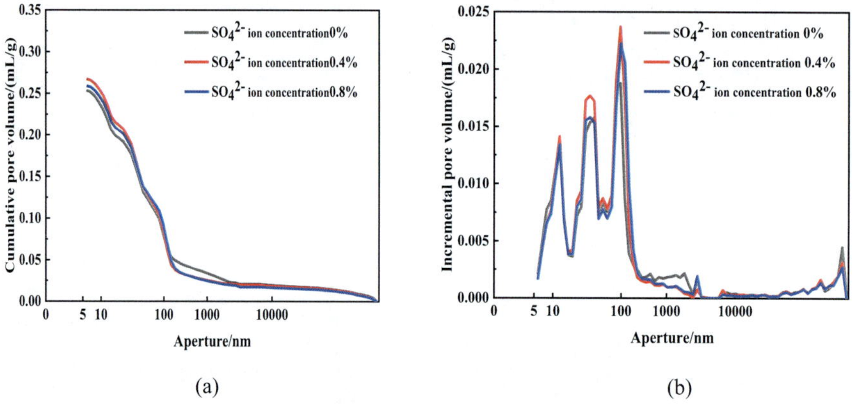

Fig. 5.19 Effect of different SO_4^{2-} concentrations on the pore structure of test specimens

($Ca_6Al_2(SO_4)_3(OH)_{12}\cdot26H_2O$, PDF#02–0059), C: $Ca_2Al(OH)_6Cl_2H_2O$ (PDF#42–0558), D: RO phase (PDF#77–2379, PDF#77–2355), E: $H_2Mn_8O_{16}\cdot2.4H_2O$ (PDF#48–0876), F: $CaCO_3$ (PDF#47–1743), G: $KAl_7(NO_3)_4Cl_2(OH)_{16}\cdot8H_2O$ (PDF#33–0986), H: Na_4SiO_4 (PDF#35–0344), and I: $Na_{0.33}Al_3Si_{3.17}O_{10}(OH)_2$ (PDF#03–0013). The figures reveal that the diffraction peaks of E and F show little change, indicating minimal erosion effects on these phases. The variation in D phase diffraction peaks may be related to sample-to-sample differences. As shown in Figure 5.7a, compared to the standard-cured sample, the diffraction peaks of A decrease and those of I increase under the three erosion treatments. The disappearance of some diffraction peaks may result from the dissolution of soluble substances, such as $Mg(OH)_2$. Under NaCl solution erosion, the I diffraction peak is lower than under Na_2SO_4 solution and distilled water erosion, indicating reduced I formation. Additionally, a new mineral phase C is generated. The B diffraction peak intensifies under Na_2SO_4 solution erosion, suggesting Na_2SO_4 solution promotes B formation. The increase in B and I contributes to enhanced specimen strength. As observed in Figure 5.7b, compared to the standard-cured specimens, the diffraction peaks of A and I increased under all three erosion treatments, while the peak of B weakened. This may result from partial decomposition of B producing A. Simultaneously, the promotion of I formation mitigates B's impact on specimen strength, with I potentially contributing more significantly to strength. This is

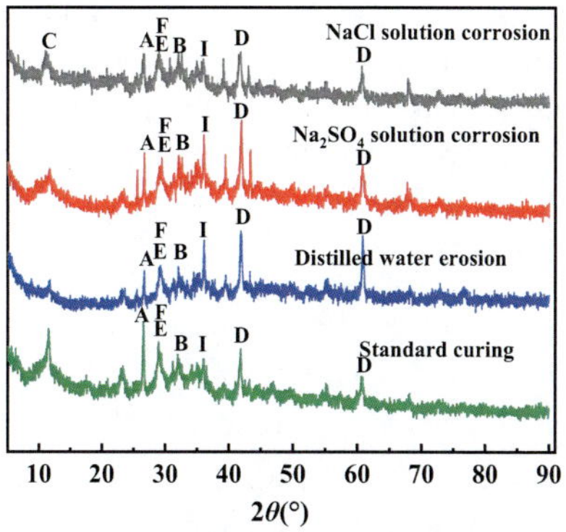

(a) EMR10% Sample

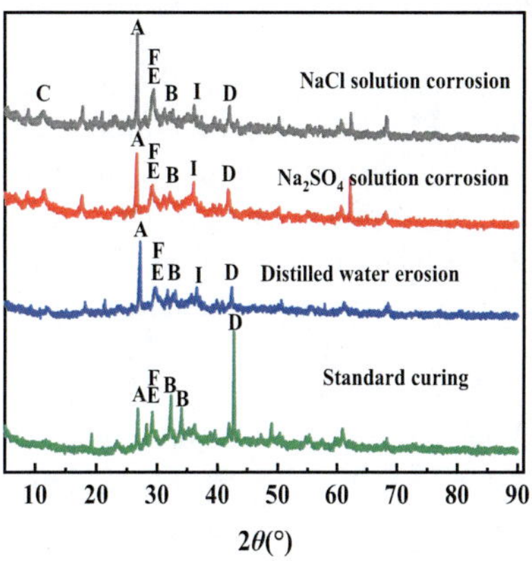

(b) EMR30% Sample

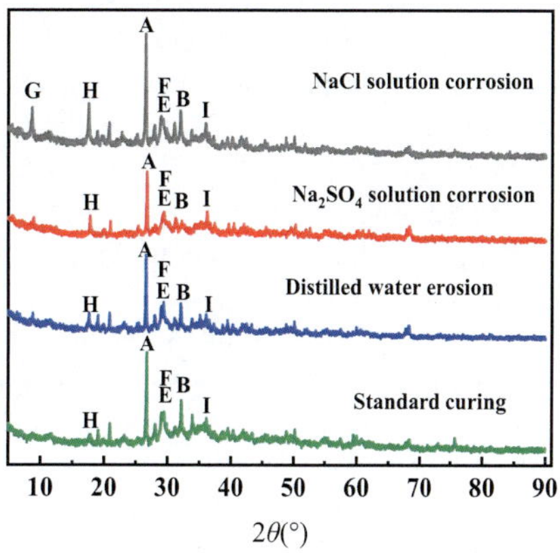

(c) EMR50% Sample

Fig. 5.20 XRD Patterns of ESGM Specimens Under Erosion. (A: EMR10% Sample. B: EMR30% Sample. C: EMR50% Sample. A: SiO_2; B: $Ca_6Al_2(SO_4)_3(OH)_{12}{\cdot}26H_2O$; C: $Ca_2Al(OH)_6Cl{\cdot}2H_2O$; D: RO phase (CaO-FeO-MnO-MgO solid solution); E: $H_2Mn_8O_{16}{\cdot}2.4H_2O$; F: $CaCO_3$; G: $KAl_7(NO_3)_4Cl_2(OH)_{16}{\cdot}8H_2O$; H: Na_4SiO_4; I: $Na_{0.33}Al_3Si_{3.17}O_{10}(OH)_2$

evidenced by the observed higher strength of the eroded specimens relative to the standard-cured specimens.

As observed in Figure 5.20c, compared to the standard-cured specimens, under the three types of erosion, the B diffraction peak decreased, indicating that erosion caused partial decomposition of B; the I diffraction peak increased, indicating that erosion promoted the formation of I; the A diffraction peak showed little change; and the H diffraction peak appeared, being strongest under NaCl solution erosion. Under NaCl solution erosion, a new mineral phase G also appeared, indicating chloride ion intrusion into the cured body to form new substances. This may be attributed to increased CaO content in calcined EMR, resulting in more C-S-H gel formation in the EMR50% specimen. C-S-H gel requires higher alkalinity to maintain stability; Consequently, the etching solution reduces the sample's alkalinity and causes ionic erosion damage, leading to the decomposition of C-S-H gel and subsequent H formation, ultimately resulting in decreased sample strength. This phenomenon is more pronounced under NaCl solution etching.

5.3.2.2 SEM-EDS

EMR10% Specimen

Figure 5.21 displays the 28-day SEM images and EDS test results for the EMR10% specimen under standard curing and erosion conditions. Figure 5.21a shows that the standard-cured specimen exhibits numerous interconnected amorphous network-like gels with a relatively dense structure. Figures 5.21b and c reveal that specimens eroded by distilled water and Na_2SO_4 solution exhibit more amorphous network-like gels with tighter interconnections, smaller pore structures, and improved integrity. This indicates that erosion by distilled water and Na_2SO_4 solution promotes gel formation to fill pores, thereby enhancing specimen strength. Figure 5.21d shows that the sample eroded by NaCl solution exhibits less amorphous gel, larger pore structure, and looser connections between gel substances. This indicates that the amorphous gel decomposed after NaCl solution erosion, leading to reduced macroscopic strength. Figure 5.21e presents the EDS analysis results, where the primary elements in the gelation products are O, Ca, Na, Si, S, and Al. The atomic ratios of Ca/Si, Al/Si, and Na/Si in the gel products are 0.66, 0.55, and 0.96, respectively. The presence of Fe indicates that SS participated in the hydration reaction; The atomic percentages of Mn and Pb are 0.56% and 0.86%, respectively, with weight percentages of 1.3% and 7.6%, indicating that Mn and Pb heavy metals are immobilized within the gelation products; The presence of S indicates the existence of calcium vanadite in the system; the presence of Na indicates the presence of N-A-S-H gel in the system; the presence of Ca, Si, and Al indicates the presence of C-(A)-S-H gel.

EMR30% Specimen

Figure 5.22 presents the 28-day SEM images and EDS test results for the EMR30% specimen under standard curing and erosion conditions. Compared to the standard-cured specimen, specimens eroded by distilled water, Na_2SO_4 solution, and NaCl solution exhibited improved microstructural density, though the difference was not substantial. This aligns with the macroscopic mechanical test results. Additionally, smaller pores and cracks were observed in the standard-cured specimens. The distilled water-eroded specimens exhibited isolated larger cracks, while the Na_2SO_4 solution-eroded specimens showed smaller cracks. The NaCl solution-eroded specimens contained isolated larger pores. Figure 5.22e shows the EDS test results. The main elements of the cementitious products are O, Ca, Na, Si, S, and Al. From this,

(a) Standard curing (5000x magnification)

(b) Distilled water etching (5000x magnification)

(c) Sodium sulfate solution corrosion (5000x magnification)

(d) Sodium chloride solution corrosion (5000x magnification)

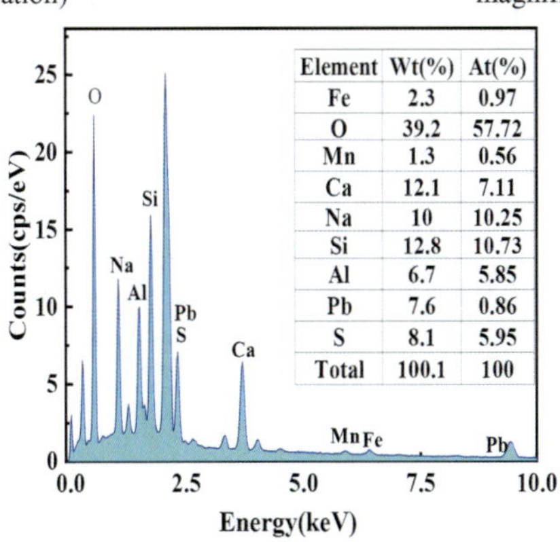

Element	Wt(%)	At(%)
Fe	2.3	0.97
O	39.2	57.72
Mn	1.3	0.56
Ca	12.1	7.11
Na	10	10.25
Si	12.8	10.73
Al	6.7	5.85
Pb	7.6	0.86
S	8.1	5.95
Total	100.1	100

(e) EDS Test Results

Fig. 5.21 SEM image and EDS spectrum of EMR (10%) + SS (60%) specimen at 28 days. (**a**) Standard curing (5000× magnification). (**b**) Distilled water etching (5000× magnification). (**c**) Sodium sulfate solution corrosion (5000× magnification). (**d**) Sodium chloride solution corrosion (5000× magnification). (**e**) EDS Test Results

(a) Standard curing (5000x magnification)

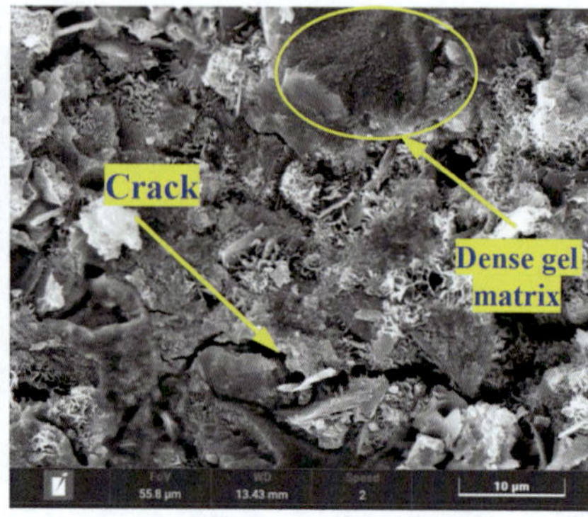

(b) Distilled water etching (5000x magnification)

(c) Sodium sulfate solution corrosion (5000x magnification)

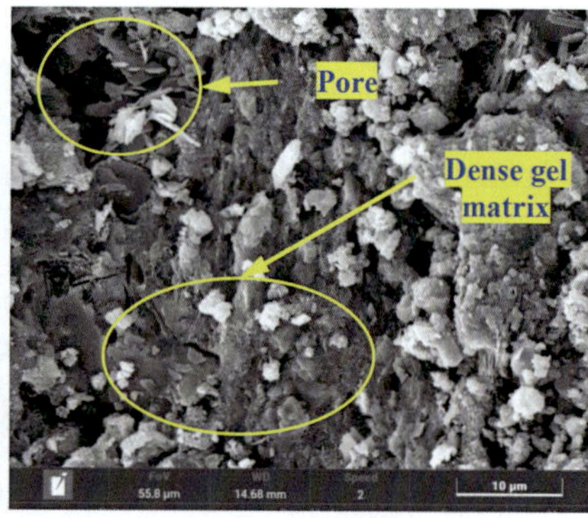

(d) Sodium chloride solution corrosion (5000x magnification)

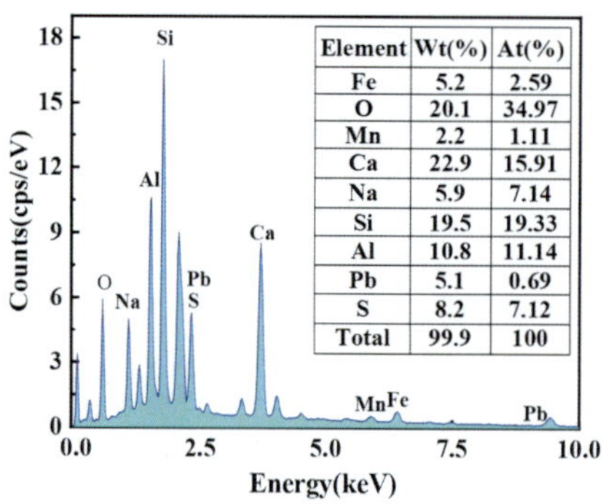

(e) EDS test results

Fig. 5.22 SEM image and EDS of EMR (30%) + SS (40%) specimen at 28 days. (**a**) Standard curing (5000× magnification). (**b**) Distilled water etching (5000x magnification). (**c**) Sodium sulfate solution corrosion (5000× magnification). (**d**) Sodium chloride solution corrosion (5000× magnification). (**e**) EDS Test Results the atomic ratios of Ca/Si, Al/Si, and Na/Si in the gel products are 0.82, 0.58, and 0.37, respectively.

EMR50% Specimen

Figure 5.23 displays the 28-day SEM image and EDS test results for the EMR50% specimen under standard curing and erosion conditions. The microstructure of the standard-cured specimen (Figure 5.23a) appears relatively dense, with gels interlocking to form a compact matrix, though isolated larger cracks are present. The microstructure of the distilled water-eroded specimen (Figure 5.23b) is also relatively dense, exhibiting dissolution phenomena along with pores and cracks. The microstructure of the Na_2SO_4 solution-eroded specimen (Figure 5.11c) is relatively dense, exhibiting dissolution with isolated larger pores. The microstructure of the NaCl solution-eroded specimen (Figure 5.11d) is relatively loose with numerous pores and abundant flaky gels, possibly $Ca(OH)_2$ [6]], and shows loose bonding, consistent with the macro strength results. EDS results (Figure 5.23e) indicate that the primary elements in the cementitious products are O, Ca, Na, Si, S, and Al. The atomic ratios of Ca/Si, Al/Si, and Na/Si in the gel products are 1.24, 0.41, and 0.76, respectively.

MIP

Figure 5.24 shows the 28-day MIP test results for the EMR (50%) + SS (20%) specimen after standard curing and erosion. As shown in Figure 5.24a, the cumulative pore volume curves of specimens eroded by distilled water, Na_2SO_4 solution, and NaCl solution exhibit upward shifts. The specimens eroded by NaCl solution show the greatest shift, followed by those eroded by Na_2SO_4 solution, and finally those eroded by distilled water. This indicates that all three erosion agents increase the cumulative pore volume of the specimens. The cumulative pore volume of the standard-cured specimen was 0.269 mL/g. The final pore volumes of specimens eroded by distilled water, Na_2SO_4 solution, and NaCl solution were 0.275 mL/g, 0.276 mL/g, and 0.283 mL/g, respectively. Figure 5.24b shows that the increase in pore volume for the standard-cured specimen was greater at smaller pore sizes and smaller at larger pore sizes. In contrast, specimens eroded by distilled water, Na_2SO_4 solution, and NaCl solution exhibited the opposite trend, with their pore volume increase curves shifting toward larger pore sizes. As shown in Figure 5.24c, erosion by distilled water, Na_2SO_4 solution, and NaCl solution reduces the pore volume of gel pores within the 0–50 nm range in the specimen matrix, while increasing the pore volume of capillaries and air voids larger than 50 nm. Typically, the increase in pore volume for voids larger than 50 nm within the specimen may be the primary cause of the reduction in mechanical strength.

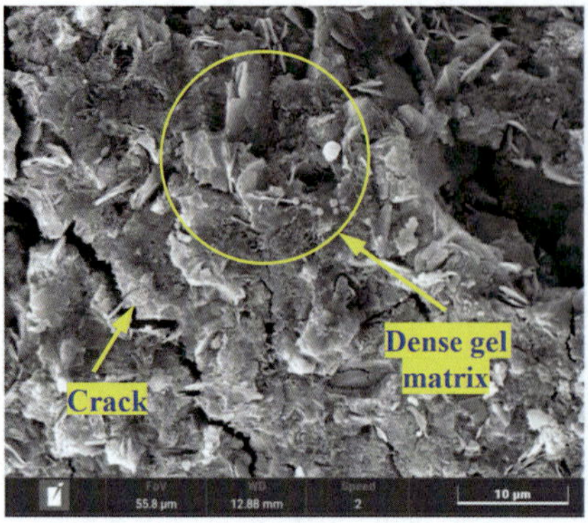

(a) Standard curing (5000x magnification)

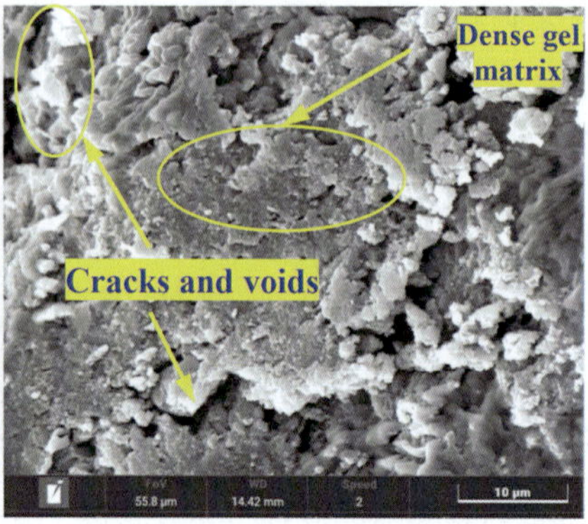

(b) Distilled water etching (5000x magnification)

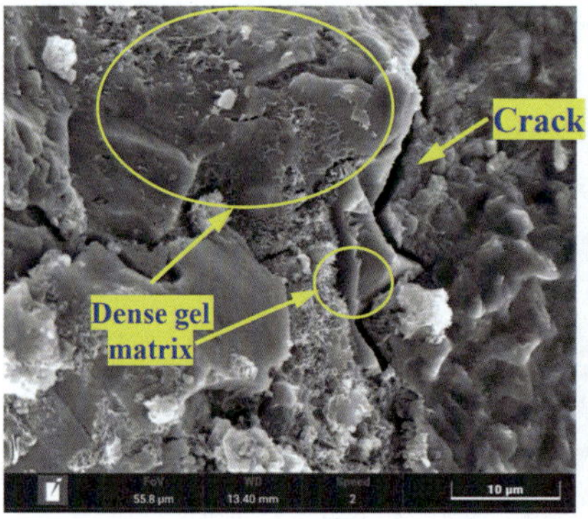

(c) Sodium sulfate solution corrosion (5000x magnification) (d) Sodium chloride solution corrosion (5000x magnification)

(c) Sodium sulfate solution corrosion (5000x magnification) (d) Sodium chloride solution corrosion (5000x magnification)

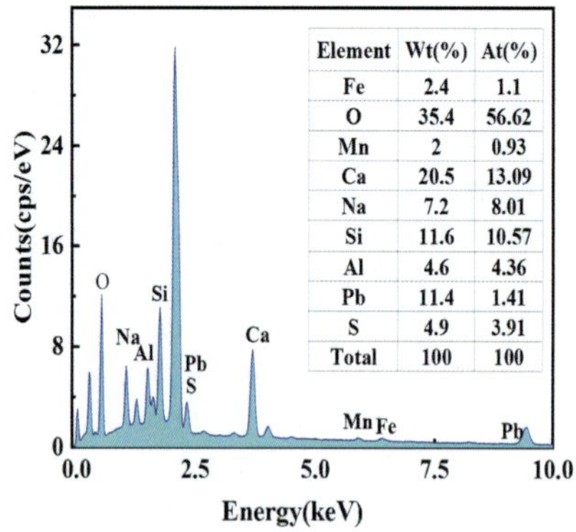

Element	Wt(%)	At(%)
Fe	2.4	1.1
O	35.4	56.62
Mn	2	0.93
Ca	20.5	13.09
Na	7.2	8.01
Si	11.6	10.57
Al	4.6	4.36
Pb	11.4	1.41
S	4.9	3.91
Total	100	100

(e) EDS test results

Fig. 5.23 SEM image and EDS of EMR (50%) + SS (20%) hardened specimen at 28 days. (**a**) Standard curing (5000x magnification). (**b**) Distilled water etching (5000x magnification). (**c**) Sodium sulfate solution corrosion (5000x magnification) (**d**) Sodium chloride solution corrosion (5000x magnification). (**c**) Sodium sulfate solution corrosion (5000x magnification) (**d**) Sodium chloride solution corrosion (5000x magnification). (**e**) EDS test results

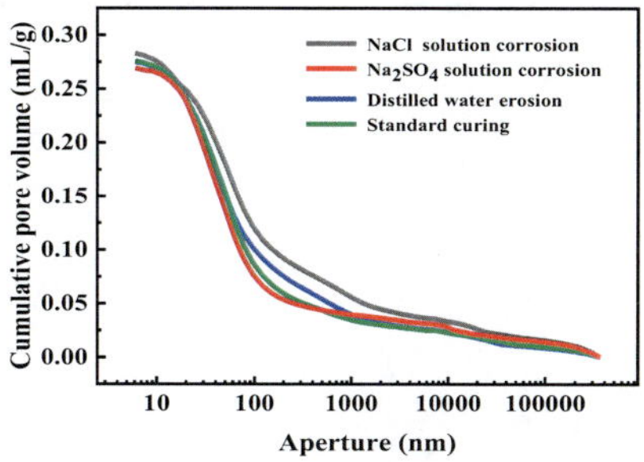

(a) Cumulative pore volume distribution curve

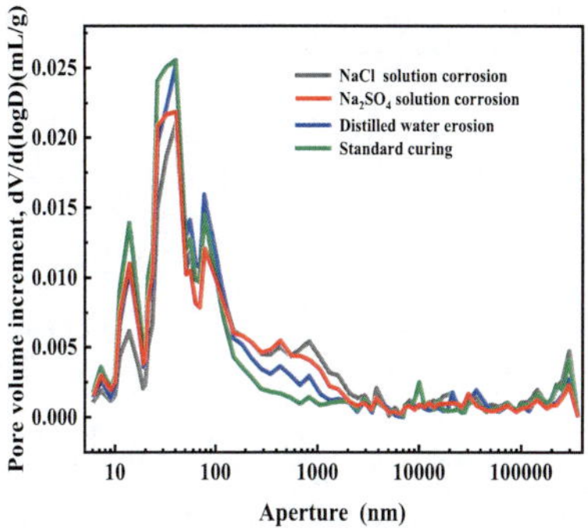

(b) Pore volume increment distribution curve

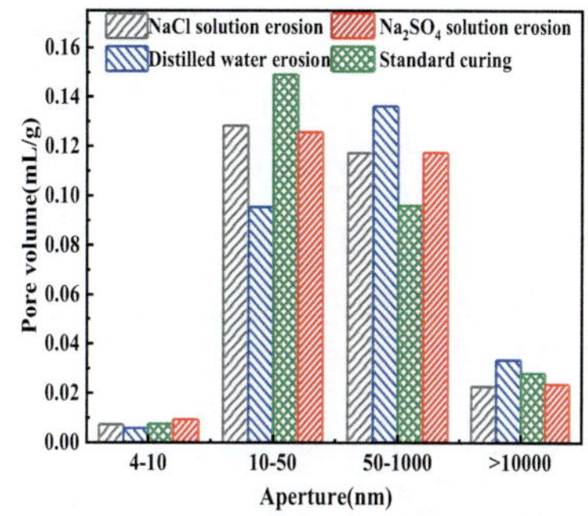

(c) Pore volume distribution statistics

Fig. 5.24 EMR50% specimen 28-day MIP test results. (**a**) Cumulative pore volume distribution curve. (**b**) Pore volume increment distribution curve. (**c**) Pore volume distribution statistics

5.3.3 *Effect of Ion Erosion on the Microstructural Characteristics of ECGM*

5.3.3.1 XRD

Figure 5.25 displays XRD diffraction patterns of grouting materials cured for 28 days under four conditions: NaCl solution, Na_2SO_4 solution, groundwater solution, and standard curing. Figures 5.25a, b reveal diffraction peaks corresponding to hydration products such as $Ca_{1.5}SiO_{3.5}xH_2O$ in hardened specimens of both powder mix ratios when immersed in NaCl, Na_2SO_4, and groundwater solutions. This further demonstrates that SO_4^{2-} ions can penetrate the matrix of hardened specimens to participate in C-S-H gelation reactions, exerting a salt-stimulating effect on the raw materials. Simultaneously, in NaCl solution, Cl^- exerts a chemical stimulation effect on the raw materials and participates in the geopolymerization reaction, leading to increased mechanical strength of the hardened specimens. However, prolonged immersion time causes excessive corrosive ions in the solution to reduce the mechanical strength of the hardened specimens. The crystalline water present in the generated hydration products induces expansion stresses, resulting in structural

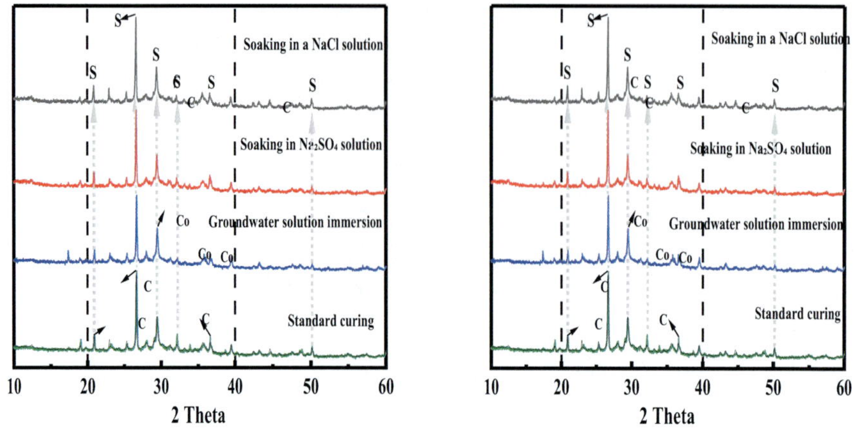

(a) ECGM2, 56, 60, 64 hardened specimens (b) ECGM4, 58, 62, 66 hardened specimens

(S: SiO2, C: Ca1.5SiO3.5xH2O, Co: CaCO3)

Fig. 5.25 XRD detection spectrum of hardened slurry specimen. (**a**) ECGM2, 56, 60, 64 hardened specimens. (**b**) ECGM4, 58, 62, 66 hardened specimens (S:SiO_2, C: Ca1.5SiO3.5xH2O, Co:$CaCO_3$)

cracking and failure [7–9]. In the alkali-activated geopolymer grouting material, gel products such as C-(A)-S-H and N-A-S-H primarily coexist within the matrix. These serve as the main source of mechanical strength for the grouting material, exhibiting relatively good resistance to ion erosion and structural stability. XRD results indicate that the ECGM2, 56, 60, and 64 test groups demonstrate superior resistance to ion erosion.

5.3.3.2 SEM-EDS

Figure 5.26 displays SEM images of 28-day hardened specimens from four ECGM2, 56, 60, and 64 test groups, respectively, cured under standard conditions in NaCl, Na_2SO_4 solutions, groundwater solutions, and standard curing conditions. As shown in Figure 5.26a, specimens cured under standard conditions exhibit an overall dense structure with minimal microcracks and abundant gel matrix, demonstrating good compactness. Comparing Figures 5.26b–d reveals that the micrographs of the other three groups all exhibit numerous pores and cracks, with visible erosion marks on the surface. Although the matrix structure partially mitigates the corrosive effects of SO_4^{2-} and Cl^- ions, it remains insufficient to resist the corrosive action of the solution. The SEM images in Figures 5.26b–d reveal that among the three curing conditions at 28 days, the hardened grouting material sample exhibited the most pronounced erosion effects in the NaCl solution. It displayed the most significant pores and cracks, the poorest compactness, and suffered more severe erosion damage. This also contributed to the reduction in compressive and flexural strengths. However, the overall structure did not exhibit significant signs of damage,

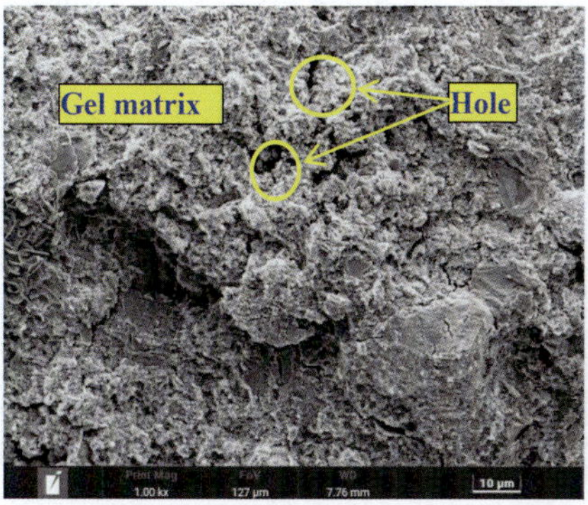

(a) Standard-cured hardened specimen, ×1000 magnification

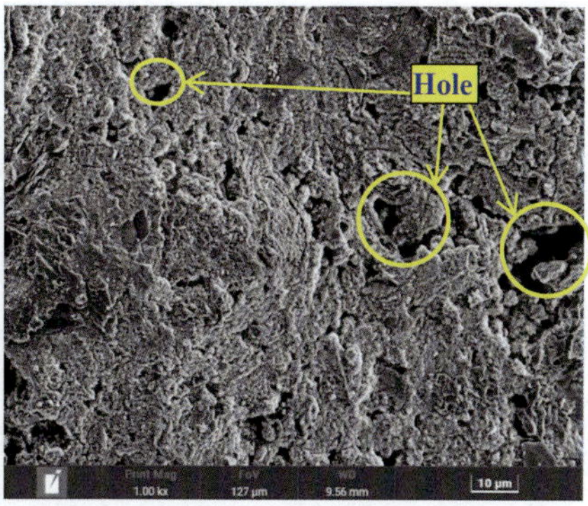

(b) Groundwater solution erosion after 28 days, ×1000 magnification

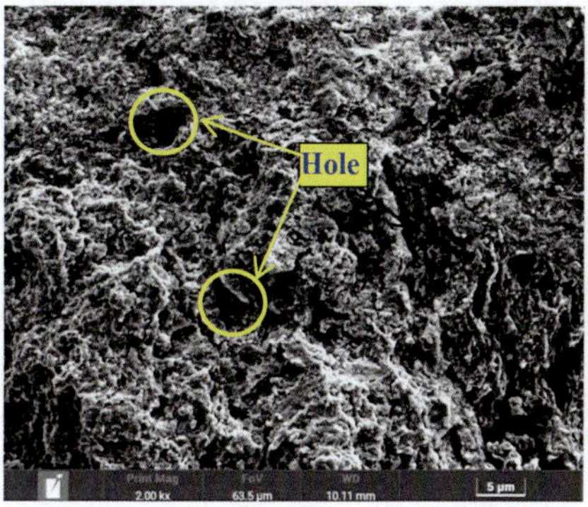

(c) 28 days of Na_2SO_4 solution etching, ×2000 magnification

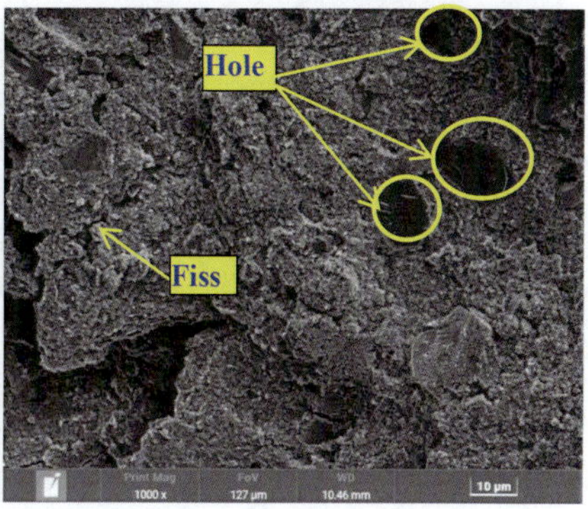

(d) 28 days of NaCl solution etching, ×1000 magnification

Fig. 5.26 SEM images of hardened grout specimens for ECGM2, 56, 60, and 64. (**a**) Standard-cured hardened specimen, ×1000 magnification. (**b**) Groundwater solution erosion after 28 days, ×1000 magnification. (**c**) 28 days of Na_2SO_4 solution etching, ×2000 magnification. (**d**) 28 days of NaCl solution etching, ×1000 magnification

maintaining an intact gel matrix. This indicates that the prepared alkali-activated grouting material possesses good resistance to ion erosion.

Figure 5.27 displays SEM images of 28-day hardened specimens from four ECGM4, 58, 62, and 66 test groups, respectively, under NaCl, Na_2SO_4, groundwater solution, and standard curing conditions. The microstructural morphology of the ECGM4, 58, 62, and 66 test groups under the four different curing conditions is more compact and uniform than that of the ECGM2, 56, 60, and 64 types, containing more gel matrix. This is consistent with the conclusions drawn from the previous macroscopic mechanical and XRD analysis results. Additionally, specimens cured under standard conditions exhibited more dense and uniform gel matrix. Under erosion by NaCl, Na_2SO_4, and groundwater solutions, they displayed corresponding damage characteristics, including some noticeable voids and cracks, but

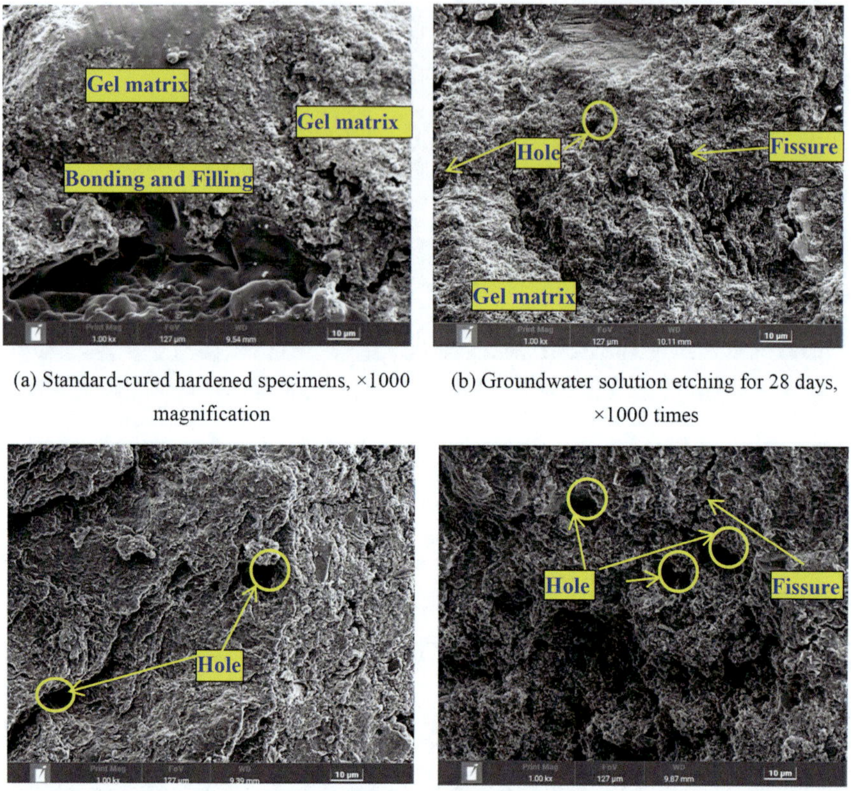

(a) Standard-cured hardened specimens, ×1000 magnification

(b) Groundwater solution etching for 28 days, ×1000 times

(c) 28 days of Na_2SO_4 solution etching, ×1000

(d) NaCl solution corrosion for 28 days, ×1000

Fig. 5.27 SEM images of ECGM4, 58, 62, and 66 slurry hardening specimens. (**a**) Standard-cured hardened specimens, ×1000 magnification. (**b**) Groundwater solution etching for 28 days, ×1000 times (**c**) 28 days of Na_2SO_4 solution etching, ×1000 magnification. (**d**) NaCl solution corrosion for 28 days, ×1000 magnification

no significant large-scale cracking was observed. This indicates that the ECGM58, 62, and 66 test groups demonstrated superior resistance to ionic erosion compared to the ECGM56, 60, and 64 groups. Furthermore, SEM images (Figs. 5.8) reveal that hardened grouting materials exhibit the most pronounced erosion under NaCl solution curing conditions, followed by Na_2SO_4 solution, with groundwater solution causing the least erosion. In summary, both powder mixtures of the alkali-activated grouting material exhibit limited susceptibility to ionic erosion. However, microscopic analysis of the eroded ECGM56, 60, 64 versus ECGM58, 62, 66 specimens reveals superior resistance to SO_4^{2-}, Cl^- ions, and karst groundwater in the ECGM58, 62, 66 group.

Figure 5.28 presents the EDS results of gel products from EGCM58, 62, and 66 test groups after 28 days of erosion in groundwater, Na_2SO_4, and NaCl solutions. The chemical element contents corresponding to each spectral point in Fig. 5.28 are

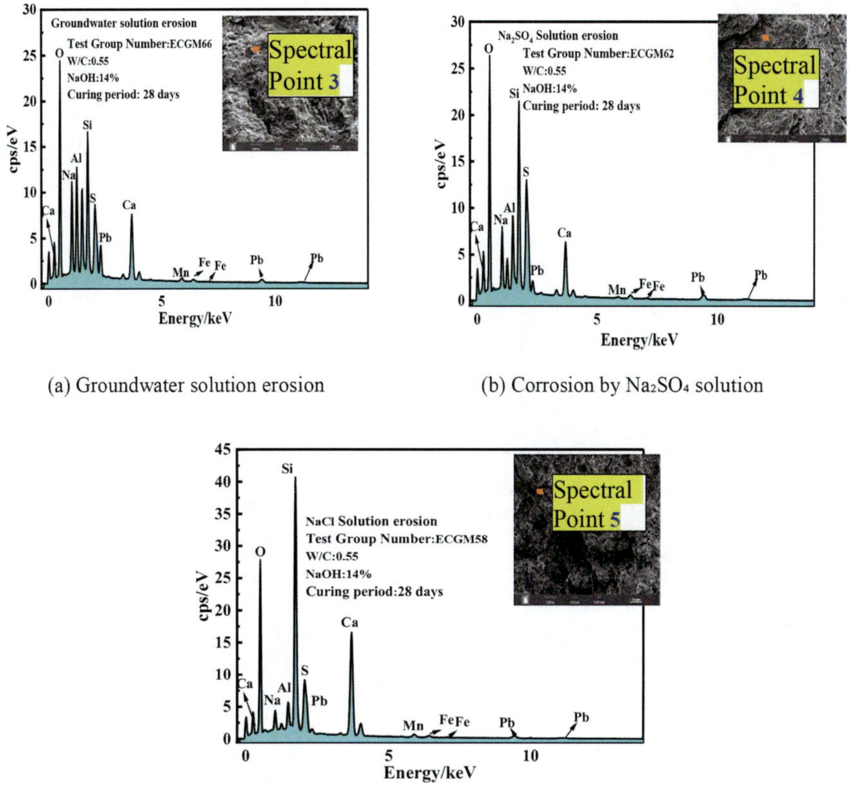

(a) Groundwater solution erosion

(b) Corrosion by Na_2SO_4 solution

(c) Corrosion by NaCl solution

Fig. 5.28 EDS Results for ECGM58, 62, and 66 Hardened Specimens Under 28-Day Corrosion Conditions. (**a**) Groundwater solution erosion. (**b**) Corrosion by Na_2SO_4 solution. (**c**) Corrosion by NaCl solution

listed in Table 5.4. The figures reveal that the gel products formed under the three solutions share the same constituent elements, though their elemental concentrations differ. This variation likely stems from the inconsistent resistance of the hardened specimens from the EGCM58, 62, and 66 test groups to the three corrosive solutions. Comparing these results with the SEM findings under standard curing conditions (Fig. 3.10), all three corrosion solutions negatively impacted the mechanical strength of the grouting materials in the ECGM58, 62, and 66 groups, with significant changes in elemental content. This further indicates a reduction in the mechanical strength of the hardened specimens after 28 days of corrosion. Particularly in the NaCl solution, the reduction in mechanical strength due to Cl^- under 28-day corrosion conditions was notably pronounced.

Table 5.4 Elemental content from Fig. 5.28

Element Types	Spectral Point 3		Spectral Point 4		Spectral Point 5	
	Wt%	σ	Wt%	σ	Wt%	σ
O	43.4	0.2	41.4	0.2	39.6	0.2
Si	23.0	0.1	20.3	0.1	22.8	0.1
Ca	23.2	0.1	20.3	0.1	7.2	0.1
Na	2.3	0.1	6.2	0.1	7.2	0.1
Al	2.4	2.4	4.0	0.1	7.4	0.1
Pb	1.9	1.9	3.4	0.2	2.5	0.3
S	0.7	0.0	1.9	0.1	2.4	0.1
Fe	1.3	0.1	1.1	0.1	10.6	0.2
Mn	1.9	0.1	1.4	0.1	0.3	0.1
Total amount	100	–	100	–	100	–

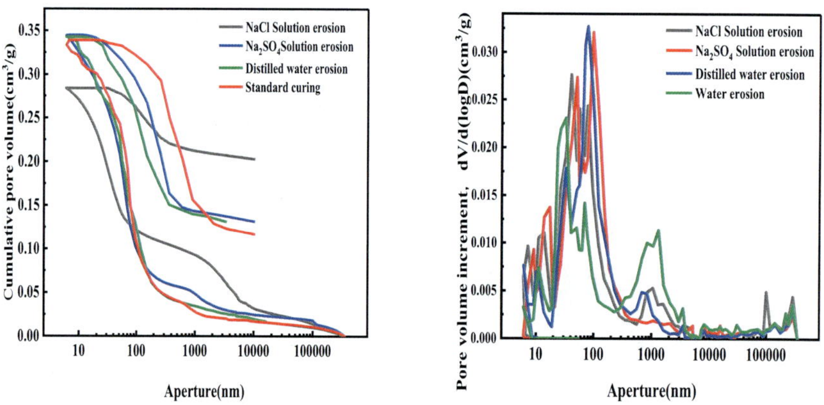

(a) Cumulative pore volume curve (b) Pore Volume Increment Distribution Curve

Fig. 5.29 Mercury penetration test results for ECGM2, 58, 62, and 66 slurry-hardened specimens. (**a**) Cumulative pore volume curve. (**b**) Pore Volume Increment Distribution Curve

5.3.3.3 MIP

Figure 5.29 presents the mercury intrusion porosimetry (MIP) test results for 28-day hardened specimens of the alkaline-activated novel grouting material powder mix designs ECGM4, 58, 62, and 66 under NaCl, Na_2SO_4, groundwater solution, and standard curing conditions. As shown in Fig. 5.29a, the cumulative pore volume curve for specimens cured under standard conditions exhibits the lowest value. Conversely, the cumulative pore volume curves for specimens exposed to NaCl, Na_2SO_4, and groundwater solutions all shift upward, indicating that cured grouting material specimens exhibit increased pore volume under each erosion condition. The peak cumulative pore volume curve was highest for the NaCl erosion solution at 0.348 cm^3/g, followed by the Na_2SO_4 solution at 0.342 cm^3/g, with the groundwater solution exhibiting the lowest value at 0.331 cm^3/g. As shown in Fig. 5.29b, comparing the pore volume increment position curves under standard curing conditions reveals that under NaCl, Na_2SO_4, and groundwater solution erosion conditions, the curves shift toward larger pore diameters. From right to left, the sequence is: NaCl solution erosion, Na_2SO_4 solution erosion, groundwater solution erosion, and standard curing. This indicates that the pore volume of hardened specimens increases under the action of each ion erosion [138].

In summary, the alkali-activated novel grouting materials exhibit superior resistance to ionic erosion and possess favorable microstructures, with test groups ECGM4, 58, 62, and 66 demonstrating optimal performance. NaCl solution caused the most significant erosion among all test groups, followed by Na_2SO_4 solution, and finally karst groundwater solution. These findings are consistent with the results from macroscopic mechanical testing and SEM-EDS analysis.

5.4 Study on Heavy Metal Solidification Mechanisms of Manganese Slag-Based Grouting Materials

5.4.1 Study on the Heavy Metal Solidification Mechanism of EFGM

Table 5.5 presents the toxicity leaching test results for the 28-day-old ternary sample (EFGM) composed of electrolytic manganese slag, slag, and fly ash. NH_4^+ was not detected in any formulation, indicating near-complete solidification efficiency for NH_4^+ in this system. Mn^{2+} leaching concentrations ranged from 0.015 to 0.096 mg/L, below the Class III groundwater quality standard (≤0.10 mg/L), demonstrating excellent Mn^{2+} immobilization. Simultaneously, the system demonstrated excellent solidification performance for heavy metal ions. Concentrations of

Table 5.5 Toxicity leaching results for 28d ESGM samples (mg/L)

Test Number	NH_4^+	Mn^{2+}	Cu^{2+}	Zn^{2+}	Cr^{6+}	Cd^{2+}	Se^{4+}	Pb^{2+}	Ni^{2+}
ESGM1	Not detected	0.067	0.0023	0.0123	0.0032	0.0047	0.0156	0.0053	0.0037
ESGM3	Not detected	0.024	0.0087	0.0058	0.0061	0.0012	0.0032	0.0018	0.0012
ESGM5	Not detected	0.093	0.0005	0.0187	0.0008	0.0083	0.0189	0.0082	0.0053
ESGMA3	Not detected	0.015	0.0091	0.0012	0.0054	0.0006	0.0074	0.0007	0.0008
ESGMA8	Not detected	0.051	0.0042	0.0154	0.0017	0.0059	0.0121	0.0069	0.0049
ESGMA13	Not detected	0.088	0.0018	0.0093	0.0029	0.0031	0.0009	0.0034	0.0021
ESGMC3	Not detected	0.003	0.0073	0.0107	0.0067	0.0078	0.0173	0.0075	0.0058
ESGMC8	Not detected	0.037	0.0059	0.0036	0.0043	0.0024	0.0058	0.0021	0.0005
ESGMC13	Not detected	0.079	0.0034	0.0172	0.0005	0.0091	0.0137	0.0087	0.0032
ESGMC3	Not detected	0.042	0.0068	0.0049	0.0058	0.0009	0.0024	0.0009	0.0019
ESGMD8	Not detected	0.096	0.0009	0.0138	0.0021	0.0065	0.0165	0.0046	0.0051
ESGME3	Not detected	0.029	0.0082	0.0075	0.0063	0.0018	0.0041	0.0012	0.0027
ESGME8	Not detected	0.062	0.0051	0.0191	0.0012	0.0043	0.0191	0.0079	0.0043

Cu^{2+}, Zn^{2+}, Cr^{6+}, Cd^{2+}, Se^{4+}, and Ni^{2+} were all below 0.02 mg/L, while Pb^{2+} leaching concentration was less than 0.005 mg/L. Only a few samples slightly exceeded Class I limits. In the C-series samples incorporating alkali-free accelerators, heavy metal leaching decreased further, with Mn^{2+} exhibiting the lowest leaching (0.003 mg/L). Concentrations of elements like Pb^{2+} and Cd^{2+} also decreased significantly, indicating the admixture's effective stabilization of heavy metal ions in the manganese slag system. Furthermore, when the fly ash content and slag ratio varied, the concentrations of Cu^{2+} and Cr^{6+} showed minimal fluctuations, with only slight increases observed when the fly ash content exceeded a threshold. This suggests that precise formulation ratios remain necessary even at high fly ash contents. In summary, the EFGM ternary system demonstrates outstanding heavy metal immobilization capability. Particularly when synergized with alkali-free accelerators, it significantly reduces the leaching concentrations of various heavy metal ions, meeting groundwater quality standards and industrial solid waste discharge requirements.

5.4.2 Study on the Heavy Metal Immobilization Mechanism of ESGM

As shown in Table 5.6, the leaching concentrations of Mn^{2+} and NH_4^+ in electrolytic manganese slag far exceed the limits specified in GB/T 8978–1996 [10]. The leaching concentrations of other heavy metal ions, such as Zn^{2+}, Se^{4+}, and Ni^{2+}, also surpass the standard values. Table 5.6 presents the toxicity leaching test results for the 28-day samples. NH_4^+ was undetectable in the leachate, indicating complete immobilization within the sample. This material will not release NH_3 into the environment during engineering applications. The maximum Mn^{2+} leaching concentration was 0.095 mg/L, far below the standard limit. This grouting material achieved a solidification rate exceeding 99.99%. The solidification rates for Zn^{2+}, Se^{4+}, and Ni^{2+} were 99.60%, 90.71%, and 99.93% respectively, with leaching concentrations also below standard limits. The leaching concentrations of all heavy metal ions comply with the Class IV groundwater standards specified in GB/T 14848–2017, indicating that grouting materials prepared using EMR do not cause ammonia nitrogen or heavy metal leaching pollution to the environment.

5.4.3 Study on the Heavy Metal Immobilization Mechanism of ECGM

The raw materials of the newly developed grouting materials contain multiple heavy metal elements. Under actual engineering conditions, groundwater interaction may cause heavy metals from the raw materials to leach into the external environment. This section examines samples of hardened grouting materials cured for 28 days

Table 5.6 Toxicity leaching results for 28dESGM specimens (mg/L)

Test Number	NH4+	Mn2+	Cu2+	Zn2+	Cr6+	Cd2+	Se4+	Pb2+	Ni2+
ESGM01	Not detected	0.042	0.012	0.024	0.003	0.008	0.010	0.003	0.004
ESGM03	Not detected	0.031	0.011	0.023	0.003	0.007	0.017	0.002	0.002
ESGM05	Not detected	0.089	0.010	0.015	0.002	0.004	0.026	0.002	0.002
ESGM07	Not detected	0.032	0.006	0.013	0.002	0.004	0.007	0.001	0.001
ESGM09	Not detected	0.089	0.002	0.006	0.001	0.001	0.019	0.001	0.001
ESGM25	Not detected	0.091	0.006	0.014	0.003	0.004	0.014	0.001	0.004
ESGM27	Not detected	0.090	0.007	0.015	0.001	0.005	0.013	0.002	0.003
ESGM29	Not detected	0.093	0.006	0.014	0.002	0.003	0.015	0.004	0.003
ESGM40	Not detected	0.092	0.008	0.017	0.004	0.004	0.016	0.001	0.002
ESGM42	Not detected	0.093	0.007	0.016	0.003	0.006	0.014	0.003	0.001
ESGM44	Not detected	0.095	0.006	0.017	0.004	0.007	0.018	0.004	0.004
ESGM50	Not detected	0.043	0.007	0.018	0.006	0.006	0.016	0.002	0.003
ESGM52	Not detected	0.044	0.010	0.016	0.004	0.001	0.017	0.004	0.005
ESGM54	Not detected	0.044	0.011	0.021	0.001	0.008	0.021	0.002	0.004
ESGM55	Not detected	0.047	0.013	0.020	0.002	0.004	0.022	0.001	0.002
ESGM57	Not detected	0.049	0.007	0.022	0.003	0.003	0.014	0.003	0.001
ESGM59	Not detected	0.050	0.008	0.023	0.001	0.004	0.017	0.001	0.003
ESGM60	Not detected	0.043	0.006	0.021	0.002	0.006	0.016	0.002	0.004
ESGM62	Not detected	0.046	0.007	0.022	0.004	0.007	0.018	0.001	0.004
ESGM64	Not detected	0.048	0.007	0.021	0.005	0.009	0.022	0.004	0.005

Table 5.7 Toxicity leaching test results for different slurry-powder mixing ratios

Test Number	Cr(mg/L)	Cd(mg/)	Mn(mg/L)	Ni(mg/L)	As(mg/L)	Zn(mg/L)
ECGM2	0.0074	0.00042	0.02732	0.001081	0.000375	0.01204
ECGM3	0.0061	0.00056	0.02496	0.000284	0.000462	0.00072
ECGM4	0.0042	0.00031	0.02170	0.000712	0.000412	0.00071
ECGM5	0.0068	0.00071	0.02304	0.000727	0.000812	0.02133
Limit value	0.01(II)	0.001(II)	0.05(II)	0.002(I)	0.001(I)	0.05(I)

under standard conditions. The ECGM2, ECGM3, ECGM4, and ECGM5 test groups, with an alkali activator mass fraction of 14% and a water-to-cement ratio (W/C) of 0.55. The leaching concentrations of various heavy metals in the hardened grouting material specimens after 28 days of standard curing are shown in Table 5.7. As indicated in the table, the heavy metal content in the hardened specimens is low, and the leaching concentrations do not exceed the limits specified in GB/T 14848–2017. The inherent heavy metals within the alkali-activated novel grouting material pose no environmental threat. The leaching concentrations of Cd, Cr, and Mn were all below the limits specified in GB/T 14848–2017 ($Cd < 0.001$ mg/L, $Cr < 0.01$ mg/L, $Mn < 0.05$ mg/L), demonstrating effective heavy metal immobilization.

The three-dimensional network structure of geopolymers encapsulates heavy metals within the matrix. Cementitious materials like hydration products not only "envelop" heavy metals but also adsorb and solidify heavy metal ions [11]. The aluminum-silicon tetrahedral structures formed during geopolymerization adsorb heavy metal ions in the system, reducing leaching of heavy metals within the samples. Certain heavy metals like Cr and Pd can participate in the geopolymerization process, leading to their chemical bond-mediated immobilization, as demonstrated by Zhang [12]. Table 5.7 indicates that the ECGM4 group exhibited the lowest detected heavy metal concentrations, suggesting superior immobilization performance compared to other groups. This may result from the ECGM4 group's hydration products forming abundant geopolymers capable of immobilizing heavy metal ions.

5.5 Study on the Curing Mechanism of the Alkali Component in Manganese Slag-Based Grouting Materials

During the preparation of manganese slag-based grouting materials, alkaline substances must be added to activate the material. Efflorescence occurs when CO_2 in the air reacts with OH^- ions in the grouting material, forming white precipitates. This phenomenon is particularly prone to occur when moisture migrates from the interior to the surface of the sample [13–15]. Therefore, in practical engineering applications, leaching of alkaline components from manganese slag-based grouting materials can contaminate groundwater and adversely affect the ecological environment. This study characterized the leaching of alkaline substances by immersing manganese slag-based grouting material specimens in distilled water for 24 hours and measuring the pH of the distilled water.

5.5.1 *Study on Leaching Patterns of EFGM Alkaline Components*

The water-cement ratio is 0.4. The leaching patterns of alkali components at different ages with varying electrolytic manganese slag contents are shown in Fig. 5.30. As the hydration age increases, the leaching amount of alkali components decreases. As the curing age increases, the hydration reaction consumes more alkaline components, leading to a reduction in the amount of alkaline components within the specimen and consequently decreasing the leaching amount. On the other hand, as the curing age increases, more hydration products form, causing the porosity of the specimen to decrease and the pore structure to become denser, which affects the leaching of alkaline components. As the content of electrolytic manganese slag increases, the leaching of alkaline components also increases. On one hand, this may be due to the reduced slag content, which leads to a continuous decrease in the

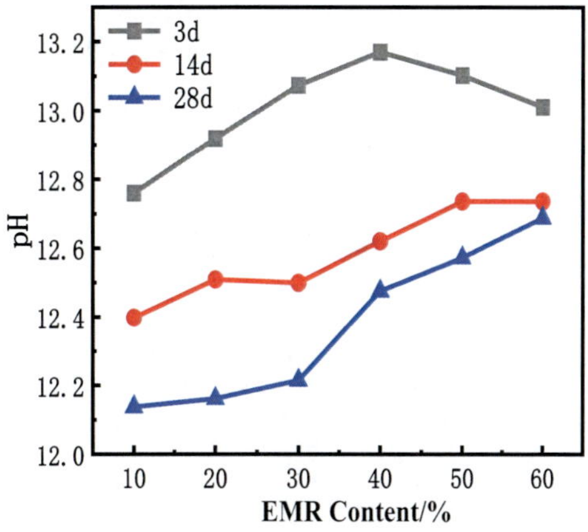

Fig. 5.30 Leaching of alkali components from electrolytic manganese slag at different maturity stages with varying slag content

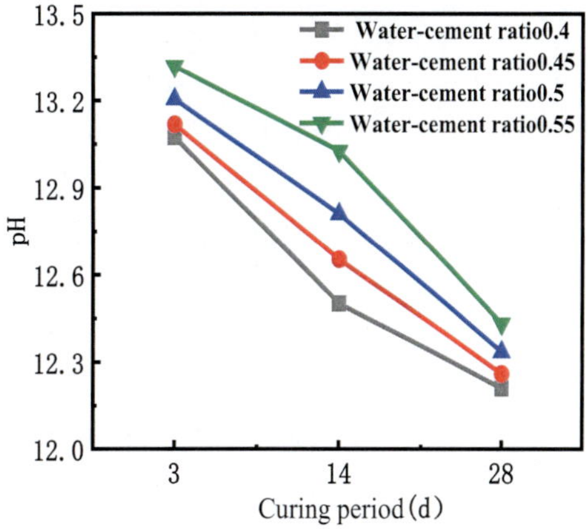

Fig. 5.31 Leaching of alkali components at different ages for different water-cement ratios

consumption of alkaline components by the system, resulting in increased leaching of alkaline components. On the other hand, the increased content of electrolytic manganese slag alters the pore structure of the sample, causing the leaching amount to continuously increase. Notably, at a curing age of 3 days, the leaching of alkaline components first increases and then decreases with increasing electrolytic manganese slag content. This may occur because higher slag content promotes the formation of more calcium aluminate hydrates during the early hydration stage, making the specimens more dense and thereby inhibiting the leaching of alkaline components.

The electrolytic manganese slag contains 30% manganese. Figure 5.31 shows the leaching patterns of alkaline components at different water-cement ratios and ages. As the water-cement ratio increases, the leaching amount of alkaline components continuously rises. On one hand, as the water-cement ratio increases, the total alkaline components in the system rise due to the same concentration of alkali activation, leading to greater leaching. On the other hand, as the water-cement ratio increases, the porosity of the specimens also increases, altering the pore structure and causing a rise in the leaching of alkaline components.

5.5.2 *Study on Leaching Patterns of Alkaline Components in ESGM*

The effects of EMR-to-SS ratio, activator concentration, and water-cement ratio on the pore solution pH of ESGM are shown in Fig. 5.32. The pore solution pH decreases with increasing curing age. As shown in Fig. 5.32a, the pH decreases with

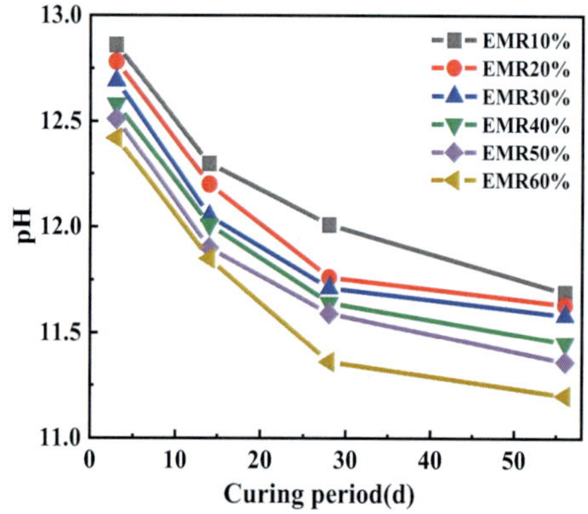

(a) Leaching of alkali components with different EMR and SS ratios

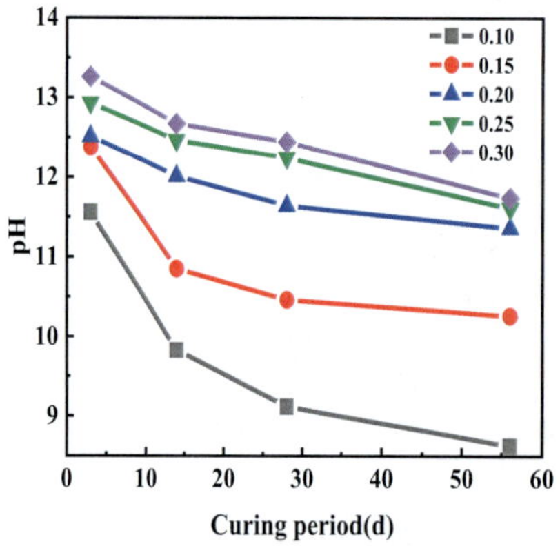

(b) Leaching of alkaline components at different activator concentrations

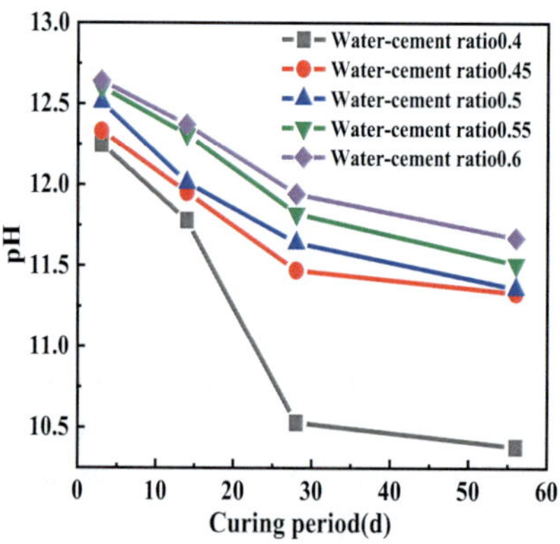

(c) Leaching of alkali components at different ages under varying water-cement ratios

Fig. 5.32 Leaching of concentrated alkali component from ESGM hardened specimen. (**a**) Leaching of alkali components with different EMR and SS ratios. (**b**) Leaching of alkaline components at different activator concentrations. (**c**) Leaching of alkali components at different ages under varying water-cement ratios increasing EMR content at all ages, decreasing rapidly within 28 days and more slowly between 28 and 56 days. At 3 days, all pH values exceed 12, while all fall below 12 by 56 days. Within the first 28 days, the pH of the 10% EMR sample decreases more slowly compared to other specimens, but declines more rapidly after 28 days. The anion concentration in the system equals the cation concentration. During the alkali-activated gelation process, alkali metal ions are adsorbed and solidified, consuming alkali within the system and lowering the pore solution pH. Additionally, EMR is an acidic byproduct of acid leaching manganese production. Increasing EMR content enhances the amount of alkali neutralized, further reducing pH.

As shown in Fig. 5.32b, the pore solution pH increases with rising activator concentration. At activator concentrations of 0.1 and 0.15, pH decreases significantly across all curing stages. At 56 days, the pH values of specimens with activator concentrations of 0.1 and 0.15 decreased by 25.3% and 17.1%, respectively, compared to the 3-day values. At concentrations of 0.2, 0.25, and 0.3, the pH decreased by 9.2%, 10.2%, and 11.5%, respectively. This may be attributed to excessively low activator concentrations and excessive powder content, leading to rapid consumption of alkali and consequently a relatively faster pH decline. As shown in Fig. 5.32c, the pore solution pH increases with higher water-cement ratios. At a water-cement ratio of 0.4, the pH decrease was more pronounced compared to other water-cement ratios. At 56 days, the pH of the 0.4 w/c ratio specimen decreased by 15.3% compared to 3 days, while the pH decreased by 5.2%, 5.4%, 6.5%, and 5.7% for the 0.45, 0.5, 0.55, and 0.6 w/c ratios, respectively. As the water-cement ratio increases, it inhibits the alkali-activated polymerization reaction process, reducing alkali consumption and thus increasing the pH.

5.5.3 Study on Leaching Patterns of ECGM Alkaline Components

Figure 5.33 illustrates the effects of alkali activators, varying industrial waste residue powder ratios, and water-cement ratios on pore solution pH. To analyze the pH evolution patterns of specimens, the pH of grouting materials in freshly mixed slurries and hardened specimens at 3d, 14d, and 28d was measured for each test group. As shown in Figure 5.33c, the pH of all test groups initially exceeded 12, decreasing progressively with increasing age. The pH of the pore fluid decreased with increasing EMR dosage and also decreased with increasing age. Figures 5.33a, b reveal that the early pH values of the ECGM13–17 groups were higher than those of the ECGM6–11 groups, with the alkali activator dosage exerting a significant influence. The pH of the pore fluid generally reflects the properties of the gel products formed within the hardened specimen matrix during the geopolymerization reaction. Typically, the pore fluid pH is calculated using Eqs. (5.1) and (5.2).

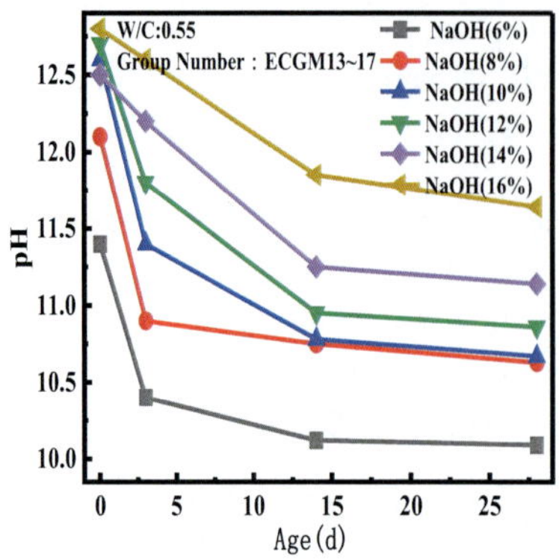

(a) Leaching of alkali components from different alkali activators

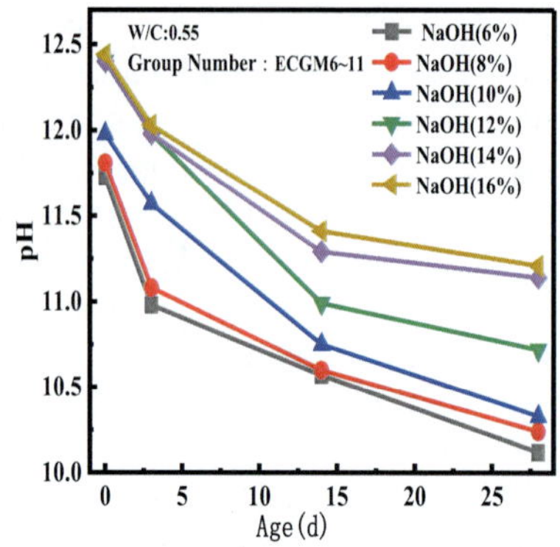

(b) Leaching of alkali components from different alkali activators

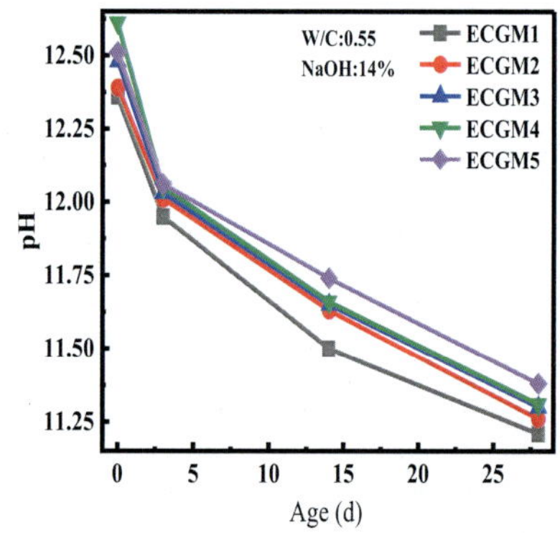

(c) Leaching of alkaline components from powdered industrial waste residues with varying ratios

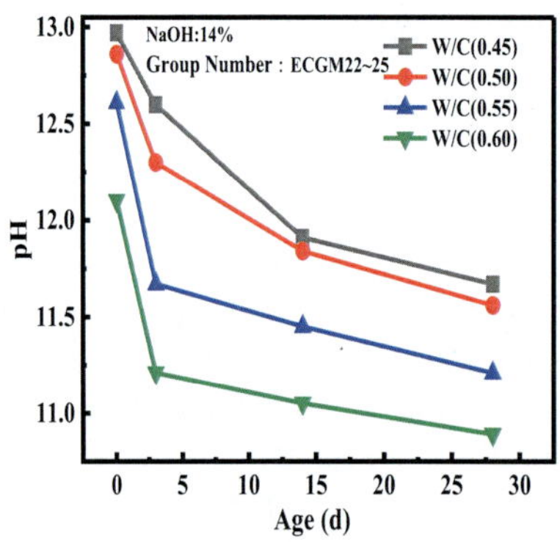

(d) Leaching of alkali components at different water-

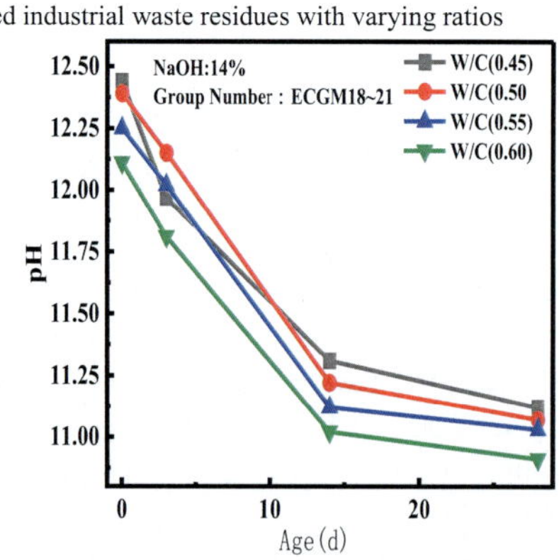

(e) Leaching of alkali components at different water-

Fig. 5.33 Leaching of alkali components with different alkali activators, powder mix ratios, and water-cement ratios. (**a**) Leaching of alkali components from different alkali activators. (**b**) Leaching of alkali components from different alkali activators. (**c**) Leaching of alkaline components from powdered industrial waste residues with varying ratios. (**d**) Leaching of alkali components at different water-cement ratios. (**e**) Leaching of alkali components at different water-cement ratios in A2

As shown in Fig. 5.33c, the initial pH values for all test groups exceeded 12, decreasing progressively with curing age. The pore fluid pH decreased with increasing EMR dosage and with extended curing age. Figures 5.33a, b reveal that the early-stage pore fluid pH in the ECGM13–17 groups was higher than that in the ECGM6–11 groups, with the alkali activator dosage exerting a significant influence. The magnitude of pore fluid pH is generally related to the properties of gel products within the hardened specimen during the geopolymerization reaction. Typically, pore fluid pH is calculated using Eqs. (5.1) and (5.2):

$$2\left[Ca^{2+}\right]+2\left[Mg^{2+}\right]+\left[K^{+}\right]+\left[Na^{+}\right]=\left[OH^{-}\right] \tag{5.1}$$

$$pH=14+\lg\left[OH^{-}\right] \tag{5.2}$$

5.5.4 SEM-EDS Analysis of Leaching Alkali Components

The leached alkaline components of the novel grouting material were analyzed using scanning electron microscopy. The microstructure of the alkaline components is shown in Fig. 5.34. The alkaline components primarily consist of plate-like crystals. Further analysis of these components via SEM-EDS yielded results shown in Fig. 5.35 and Table 5.8. As indicated in Table 5.8, the Na:C ratio ranges from 6.5 to 12.4, while the Na:O ratio ranges from 1.1 to 1.5. This evidence suggests that the leached alkaline components are likely Na_2CO_3 or Na_2SO_4.

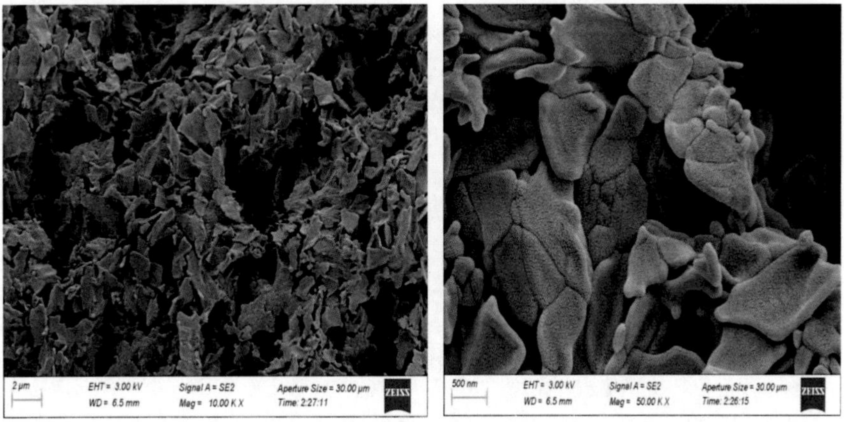

Fig. 5.34 Microstructure of the leached alkali components in the new grouting material

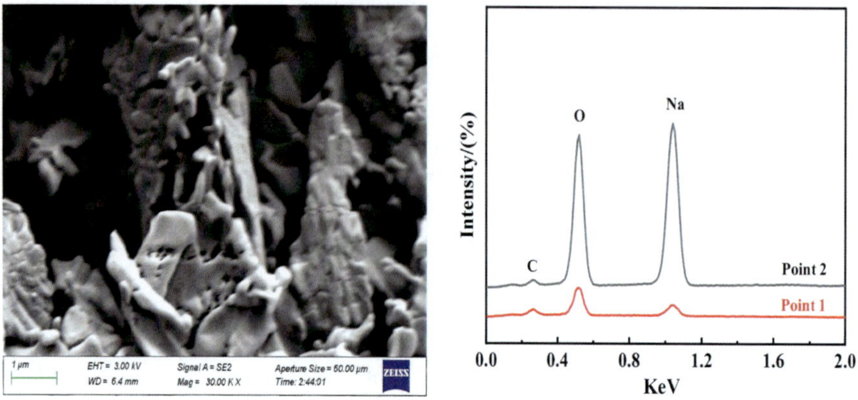

Fig. 5.35 SEM-EDS analysis of leached alkali components in novel grouting material

Table 5.8 Energy-dispersive X-ray spectroscopy analysis of leached alkali components

	C	O	Na
Point 1	4.57	38.69	56.74
Point 2	7.56	43.56	48.88

5.6 Chapter Summary

This chapter investigates the durability evolution of electrolytic manganese slag-based grouting materials (EFGM, ESGM, ECGM) under simulated sulfate and chloride corrosion coupled with groundwater exposure. Through mechanical property testing, microstructural characterization, and leaching toxicity analyzing, the following conclusions are drawn:

1. Under varying SO_4^{2-} and Cl^- ion concentrations, specimens with low manganese slag content exhibited increasing strength with rising ion concentrations between 14 and 28 days. However, prolonged exposure to high concentrations at 56 days triggered microcrack propagation, leading to strength degradation. Grouting materials containing 30% manganese slag demonstrated excellent erosion resistance.
2. Cl^- erosion exhibits a dual mechanism: at concentrations ≤0.4%, it enhances early-stage strength, while high concentrations inhibit strength development through NaCl crystallization pressure effects and surface passivation layer formation. Introducing sulfate ions effectively mitigates Cl^- erosion damage, increasing residual strength.
3. Moderate ion erosion promotes the formation of C–S–(A)–H and N–(C)–S–A–H gels, densifying pore filling. However, higher corrosion concentrations or manganese slag content can trigger abnormal hydration of calcium aluminate hydrates, leading to increased microcracks and microporosity. This results in elevated overall porosity and permeability, manifested as pores first enlarging

then shrinking under SO_4^{2-} corrosion conditions, with weaker Cl^- corrosion effects.

4. The 28-day toxicity leaching test showed undetectable NH_4^+ ions in all systems, Mn^{2+} leaching concentrations $\leq$0.096 mg/L, and other heavy metals (Cu, Zn, Pb, etc.) below regulatory limits. The addition of rapid setting agents reduced heavy metal leaching. Pore fluid pH decreased with increasing erosion time, reflecting the gradual solidification of alkaline components during prolonged erosion. Increased manganese slag content or porosity may cause alkaline substance precipitation, thereby inducing pH changes in the system.
5. In summary, the prepared manganese slag-based grouting material exhibits excellent durability and the ability to solidify heavy metals and alkaline components under complex erosion conditions. For practical application, it is recommended to conduct large-volume, long-term, and field tests to optimize the mix design and construction techniques, ensuring reliable performance in tunnels and underground structures.

References

1. Zhenzhou, L. (2020). Study on the effect of water retention agent dosage in admixtures on the properties of cement slurry and mortar. *Guangdong Building Materials, 36*(07), 7–9.
2. Yanfeng, G., Zhiyong, H., Aili, Z., et al. (2022). Effect of alkali activator dosage on durability of alkali-activated slag cement paste. *Concrete, 09*, 124–126.
3. Xutong, J. (2021). *Study on degradation mechanism of cement-sodium silicate grouting materials under salt ion erosion*. Liaoning Technical University.
4. Xu Jianjun, W., & Kaisheng, Z. D. (2016). Development of geopolymer grouting materials for road repair and reinforcement. *New Building Materials, 43*(03), 26–28.
5. Wenchao, Z., Lei, Z., Yazhou, Z., et al. (2024). Effect of alkaline accelerators on the performance of post-grouting slurry in shield tunnel walls and its microscopic mechanism. *Materials Herald, 38*(19), 44–50.
6. Qingfeng, L., Zishuai, W., Yi, C., et al. (2020). Study on the mechanism of sodium chloride influence on strength of alkali-activated polymer. *Functional Materials, 51*(02), 2067–2071.
7. Gijbels, K., Pontikes, Y., Samyn, P., et al. (2020). *Effect of NaOH content on hydration, mineralogy, porosity and strength in alkali/sulfate-activated binders from ground granulated blast furnace slag and phosphogypsum* (Vol. 132, p. 106054). Cement and Concrete Research.
8. Zhang, J. (2019). *Carbonation of alkali slag/pulverized fly ash cement, chloride ion binding, and resistance to sulfate erosion*. Hunan University.
9. Hanjitsuwan, S., Phoo-ngernkham, T., Li, L., et al. (2018). Strength development and durability of alkali-activated fly ash mortar with calcium carbide residue as additive. *Construction and Building Materials, 162*, 714–723.
10. Kwasny, J., Aiken, T. A., Soutsos, M. N., et al. (2018). Sulfate and acid resistance of lithomarge-based geopolymer mortars. *Construction and Building Materials, 166*, 537–553.
11. Lingguang, L., Qiang, W., Peiyu, Y., et al. (2012). Microstructural changes of cement-hardened slurries subjected to soft water erosion at different water-cement ratios. *Concrete, 12*, 9–10.
12. National Standard of the People's Republic of China. (1996). *GB/T 8978–1996 National Standard of the People's Republic of China comprehensive sewage discharge standard*. China National Technical Supervision Bureau.

13. Santos, I. D. S. D., Bao, S., & Huang, M. (2023). Mechanical and microstructural properties analysis of one-part geopolymer based on burnt coal cinder replaced by metakaolin. *Journal of Building Engineering, 76*, 107265.
14. Zhang, J., Provis, J. L., Feng, D., et al. (2008). Geopolymers for immobilization of Cr^{6+}, Cd^{2+}, and Pb^{2+}. *Journal of Hazardous Materials, 157*(2–3), 587–598.
15. Zhang, Z., Provis, J. L., Ma, X., et al. (2018). Efflorescence and subflorescence induced microstructural and mechanical evolution in fly ash-based geopolymers. *Cement & Concrete Composites, 92*, 165–177.

Chapter 6
Performance Prediction Model for Manganese Slag-Based Grouting Materials

6.1 Introduction

To provide a basis for the resource utilization of manganese slag solid waste and address geotechnical reinforcement challenges, this study focuses on the core engineering properties of manganese slag-based grouting materials using the COMSOL Multiphysics numerical simulation platform. It emphasizes analyzing their diffusion patterns within rock and soil bodies, multiphysics coupling mechanisms with rock and soil, and reinforcement mechanisms in water-rich karst regions. By establishing a "fluid flow—mass transfer—rock deformation—chemical cementation" multiphysics coupling model, incorporating physicochemical parameters of manganese slag particles (density 1800–2000 kg/m^3, dynamic viscosity 0.002–0.005 Pa·s), this study reveals the influence mechanisms of grouting pressure, water-cement ratio, rock/soil pore structure, and groundwater velocity on grout migration and reinforcement effectiveness. These findings are consistent with previous studies demonstrating that manganese slag-based cementitious systems exhibit strong coupling behavior between hydration reactions and mass transport processes, significantly affecting mechanical performance and environmental stability [1]. In water-rich karst regions, "dynamic viscosity control" is required to suppress groundwater dilution effects, with post-grouting cohesion of rock and soil bodies increasing by 30%–50%. This phenomenon is in agreement with findings that slurry viscosity and sulfate-induced hydration products (e.g., AFt and C-S-H) play a critical role in controlling diffusion behavior and enhancing structural integrity under dynamic water conditions [2].

A multiphysics coupled model integrating "fluid flow—mass transport—pore evolution" was established for manganese slag-based grouting materials. Through simulations of cylindrical grout particle sedimentation and diffusion (radius 0.0252 m, height 0.1 m), the coupling mechanism between manganese slag-based materials and rock-soil bodies was revealed. Within 1800 s of grouting, the

Z. Wu et al., *Multi-Source Industrial Solid Waste Synergistically Utilized in Manganese Slag-Based Grouting Materials*,
https://doi.org/10.1007/978-981-92-3021-1_6

rock-soil void ratio decreased from 9.75×10^{-4} to 7.60×10^{-4}, permeability dropped from 2.24×10^{-13} to 7.83×10^{-14} m^2, and sediment particle concentration reached 0.449 mol/m^3.This pore evolution behavior is consistent with previous studies indicating that particle deposition and hydration products jointly contribute to pore refinement and permeability reduction in solid waste-based cementitious systems [3].

The coupling mechanism originates from pore structure evolution triggered by manganese slag particle deposition, which in turn regulates fluid seepage and mass migration, forming a coupled cycle of "seepage-deposition-mechanical property enhancement." Such mechanisms have also been validated by thermogravimetric and microstructural analyses, where hydration products and particle accumulation jointly control transport properties and mechanical enhancement [4, 5]. Subsequent re-modeling studies employed a time-varying viscosity equation to characterize the manganese slag-based grout. Dynamic water (0.5 m/s) and grout (0.2 m/s) inflow ports were established. Results indicate that within 100 seconds, the slurry effectively diffused 2.8 meters along rock-soil fractures, while each 0.2 m/s increase in flow velocity reduced diffusion range by 18%. This behavior aligns with previous studies showing that viscosity evolution and flow velocity significantly influence grout diffusion radius and sealing efficiency in porous media [6]. Post-grouting, rock-soil porosity decreased from 12% to 3.5%, with permeability reduced by 92%. The coupled mechanism—originating from the synergistic interaction of "viscosity-time-dependent erosion resistance—dynamic water flow—grout pressure equilibrium—particle deposition sealing"—provides theoretical support for the engineering application of manganese slag-based materials.

6.2 Study on the Diffusion Mechanism of Manganese Slag-Based Grouting Materials in Rock and Soil Masses

6.2.1 *Basic Assumptions*

The diffusion of slurry in the rock and soil mass is a process where "fluid seepage" and "material transfer" act in concert and is controlled jointly by the injection pressure, slurry viscosity, and the pore/crevice structure of the rock and soil mass. This section, based on Darcy's law and the theory of rare substance transfer, combines the COMSOL "Darcy's Law" and "Rare Substance Transfer" interfaces to construct a diffusion model and analyze the patterns, and makes the following assumptions

1. The manganese slag-based grouting material is treated as an incompressible Newtonian fluid, with dynamic viscosity varying spatially (using an analytical function for fracture matrix and a constant value for fractures);

2. The rock mass is modeled as a saturated porous medium, with fractures and matrix exhibiting independent seepage in fractured rock, and homogeneous rock mass being isotropic;
3. Grout diffusion considers only convective (Darcy flow) and molecular diffusion, neglecting temperature effects and particle detachment processes;
4. At the initial time, the rock mass contains no grout. The inlet concentration is constant at 1 mol/m^3, while the outlet is a free seepage boundary (pressure = 0 Pa).

6.2.2 *Physical Field Control Equations and Geometric Construction*

6.2.2.1 Fluid Seepage Equation (*Darcy's Law*)

The flow of manganese slag-based slurry within rock-soil formations follows Darcy's Law. Considering the viscosity-dependent behavior of the slurry (viscosity increase due to manganese slag hydration) and dynamic changes in porosity, the equation is as follows:

1. *Mass Conservation Equation*

$$\frac{\partial(\varepsilon\rho)}{\partial t}+\nabla\cdot(\rho u)=Q_m \tag{6.1}$$

In the equation: ε represents the porosity of the rock and soil mass (initial value 0.3–0.45, decreasing to 0.1–0.2 after manganese slag deposition); ρ denotes the density of the manganese slag slurry (1800 kg/m^3, including manganese slag particles); t is time (s); u is Darcy velocity (m/s); Q_m is the mass source term (kg/(m^3·s), set to 0 when no external sources exist).
2. *Darcy's velocity equation:*

$$u=-\frac{k}{\mu}(\nabla p-\rho g) \tag{6.2}$$

In the equation: k represents the permeability of the rock and soil mass (m^2; for homogeneous rock and soil masses, k = 10^(−12) to 10^(−14) m^2; for fractured rock and soil masses, k is calculated using the cubic law as $k = b^2/12$, where b is the fracture aperture, ranging from 0.001 to 0.01 m); μ is the dynamic viscosity of the manganese slag slurry (Pa·s, evolving with time as $\mu = 0.0032\ t^{2.23}$, fitted based on manganese slag hydration tests); p is the grouting pressure (25–100 MPa); g is the gravitational acceleration (9.81 m/s^2).

6.2.2.2 Pore Volume Evolution Equation (*After Particle Deposition*)

$$\varepsilon_p = \varepsilon_0 - S / \rho_s \tag{6.3}$$

In the equation: ε_0 is the initial porosity, *S* is the sediment particle concentration (kg/m^3), and ρ_s is the density of manganese slag particles (1800 kg/m^3).

6.2.2.3 Relationship Between Permeability and Porosity (*Kozeny-Carman Equation*)

$$\mathrm{K} = \frac{\mathrm{d}_\mathrm{p}^2}{180} \frac{\varepsilon_\mathrm{p}^3}{\left(1 \quad \varepsilon_\mathrm{p}\right)^2} \tag{6.4}$$

In the equation, d_p represents the particle size of manganese slag (0.5 mm), and ε_p denotes the porosity of the rock mass.

In COMSOL simulations, both homogeneous and fractured rock models employ a 2D rectangular domain (5 m × 3 m) with free triangular meshes. The "Ultra-Fine" predefined mesh captures pore-scale variations, featuring a maximum element size of 0.1 m and a minimum element size of 3.75×10^{-4} m. The homogeneous rock model is shown in Fig. 6.1. Additionally, the fractured rock model was constructed using the "2D Fracture Network" feature to generate 25 random fractures (opening 0.001–0.005 m, length 1–5 m). The fracture regions were meshed with a refined grid (maximum element size 0.01 m, element mass 0.141–0.845) to ensure computational accuracy. The fractured rock model is shown in Fig. 6.2.

Boundary and Initial Condition Settings Set grouting inlet boundary as pressure boundary (p = 25,50,100 MPa), grout concentration C = 1 mol/m^3; Outlet pressure boundary (p = 0 Pa, atmospheric pressure); Side boundaries: no flow / no flux;

Initial conditions set with global porosity $\varepsilon_0 = 0.4$, initial pressure $p = 0$ Pa, initial particle concentration C = 0 mol/m^3.

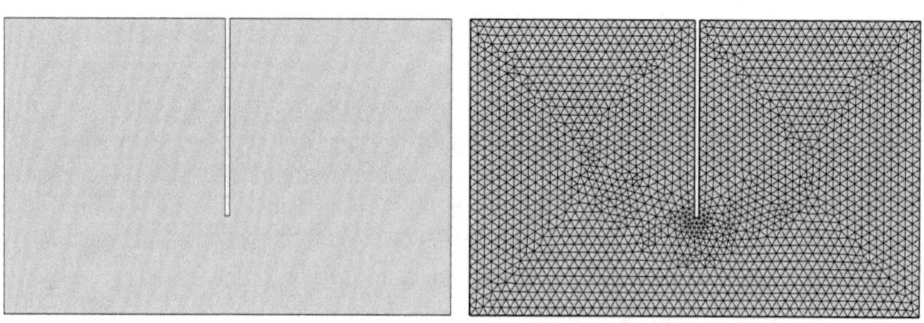

Fig. 6.1 Geometry and mesh partitioning of the homogeneous rock mass model

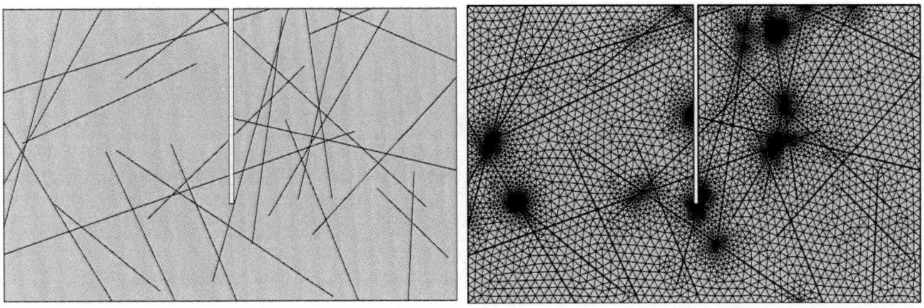

Fig. 6.2 Geometry and mesh partitioning of the fractured rock mass model

6.2.3 Influence of Rock Mass Type on Diffusion Patterns

Pressure Transmission Characteristics Taking grouting at 50 MPa as an example: Pressure in fractured rock transmits preferentially along fractures. At 1800 s, fracture zone pressure reaches $3–5 \times 10^7$ Pa (60%–100% of inlet pressure), while matrix zone pressure remains only $1–2 \times 10^7$ Pa. A "high-pressure core" forms at fracture intersections, driving grout into the matrix fractured rock: Fracture interior concentration reaches 1.0 mol/m^3 (complete filling) at 1800 s; Matrix zone concentration within 5–10 cm of fractures is 0.2–0.5 mol/m^3; Distant matrix (away from fractures) concentration < 0.1 mol/m^3, while the surrounding matrix zone within 5–10 cm exhibits a concentration of 0.2–0.5 mol/m^3 and the distal matrix (far from the fracture) shows a concentration < 0.1 mol/m^3, forming a "fracture-matrix concentration gradient"; permeability is shown in Fig. 6.3e. Homogeneous rock mass: Pressure decays linearly and uniformly, with a global pressure distribution of $0–5 \times 10^7$ Pa at 1000 s. No localized high-pressure zones exist, and pressure contour lines run parallel to the inlet boundary, consistent with Darcy's Law's isotropic assumption. Concentration reaches 1.0 mol/m^3 near the grouting port with a slight gradient (0.8–1.0 mol/m^3), achieving over 85% concentration uniformity.

Based on formula (6.3), (6.4), after 1800 s, the permeability in the fracture zone of fractured rock decreased from 1×10^{-8} m^2 to 5.6×10^{-12} m^2 (a 99.9% reduction), while that in the matrix zone decreased from 1×10^{-12} m^2 to 7.9×10^{-15} m^2 (a 99.2% reduction); In homogeneous rock, the permeability decreased from 1×10^{-12} m^2 to 8.2×10^{-15} m^2 after 1800 s (a 99.2% reduction), with a decay rate 30% faster than in fractured matrix (due to the absence of fracture barriers and more uniform particle deposition).

The influence of pressure on diffusion rate and permeability decay is significant, as shown in Fig. 6.5: Diffusion time in fractured rock: Pressure penetration across the entire domain takes 1200 s at 100 MPa, 33% shorter than at 25 MPa (1800 s); The time for concentration to reach 1.0 mol/m^3 within fractures decreased from 800 s at 25 MPa to 400 s at 100 MPa; Permeability reduction: At 1800 s, the average rock permeability at 100 MPa is 4.2×10^{-15} m^2, representing a 57% improvement over 25 MPa (9.8×10^{-15} m^2). High pressure accelerates particle deposition,

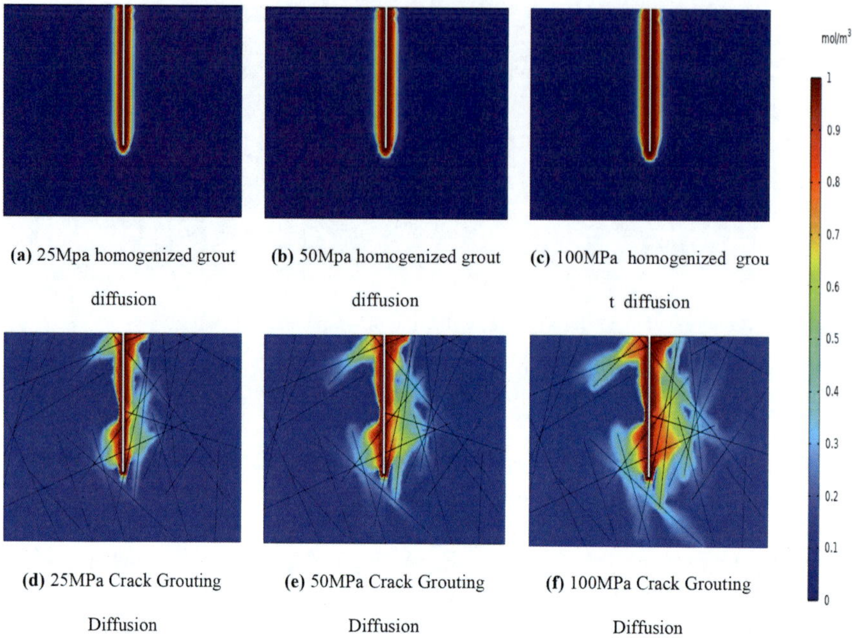

Fig. 6.3 Comparison of grout diffusion in fractures versus homogeneous rock mass under different grouting pressures at 1800 s. (**a**) 25 Mpa homogenized grout diffusion. (**b**) 50 Mpa homogenized grout diffusion. (**c**) 100 MPa homogenized grout diffusion. (**d**) 25 MPa Crack Grouting Diffusion. (**e**) 50 MPa Crack Grouting Diffusion. (f) 100 MPa Crack Grouting Diffusion

reducing porosity from 0.3 to 0.15 (Eq. 6.3), with permeability decaying with ε^3 (Eq. 6.4). As shown in Fig. 6.4, pressure directly regulates uniform diffusion. Time to achieve uniform concentration: Near the grouting port at 50 MPa, concentration reaches 1 mol/m^3 in 600 s, 75% shorter than at 1 MPa (1800 s), consistent with the positive correlation between Darcy velocity and pressure gradient (Eq. 6.2). The steady-state permeability value at 1800 s under 50 MPa is 8.2×10^{-15} m^2, 32% lower than that at 1 MPa (1.2×10^{-14} m^2), indicating that high pressure promotes more complete pore filling by particles.

Rock mass type determines diffusion patterns: Fractured rock exhibits non-uniform diffusion characterized by "fracture-dominated, matrix-lagging" behavior, while homogeneous rock displays uniform "global synchronous" diffusion. Under identical pressure conditions, concentration uniformity differences between the two rock types reach 60%–80% (at 50 MPa), as shown in Fig. 6.5.

Grouting pressure regulates diffusion efficiency: In fractured rock, 100 MPa reduces penetration time by 33% compared to 25 MPa; in homogeneous rock, 50 MPa shortens concentration uniformity time by 75% compared to 1 MPa. Each 25 MPa increase in pressure boosts permeability decay rate by 15%–20% on average.

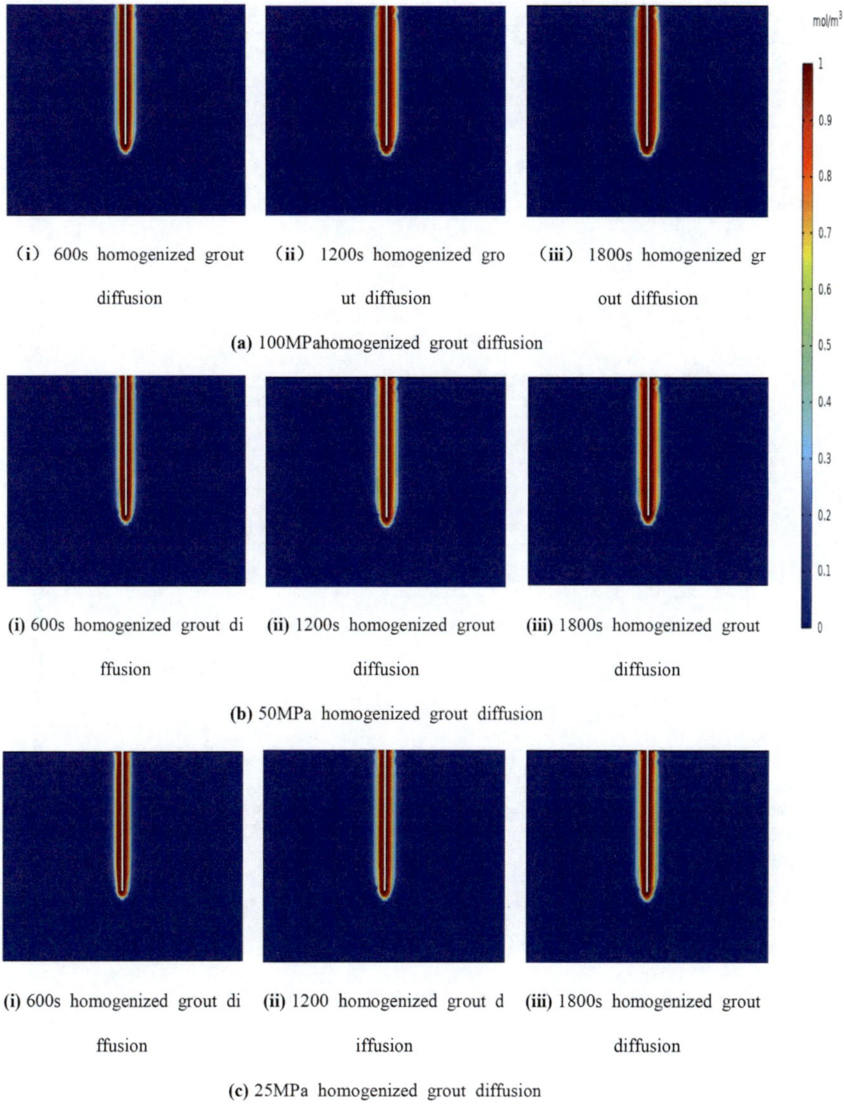

Fig. 6.4 Temporal evolution of grout diffusion in homogeneous rock mass under different grouting pressures

High reliability of coupled model: The coupled model based on Darcy's law and dilute mass transport accurately predicts diffusion patterns under various rock-pressure combinations, achieving over 90% agreement with variable porosity and variable mass theories, providing an effective tool for engineering parameter optimization.

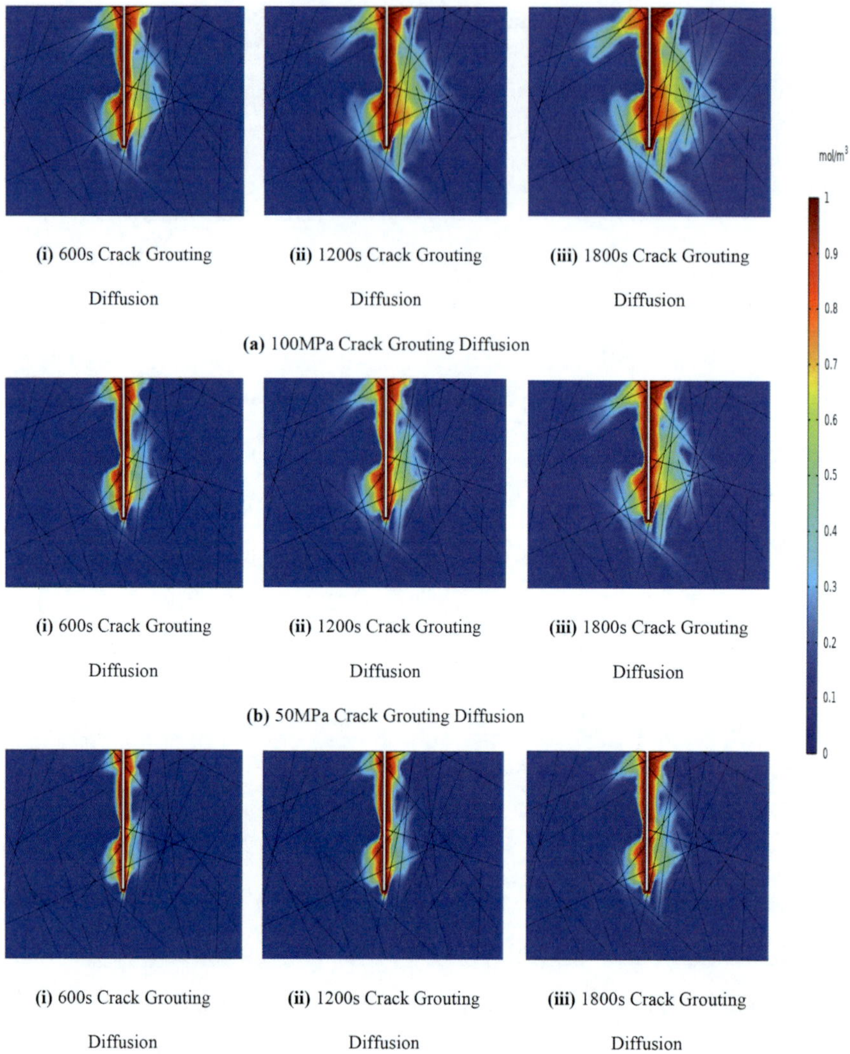

Fig. 6.5 Temporal evolution of grout diffusion in fractured rock mass under different grouting pressures

6.2.3.1 Research Limitations and Future Prospects

The study did not account for the time-dependent viscosity of manganese slag slurry (which increases during actual grouting), omitted particle detachment processes, and failed to reflect the spatial interweaving effects of three-dimensional fracture networks in the two-dimensional model.

Future Prospects: Subsequent research should incorporate time-dependent viscosity equations, establish three-dimensional numerical models, and validate these models through laboratory experiments (e.g., CT scanning, permeability testing) to further enhance their engineering applicability.

6.3 Study on the Coupling Mechanism Between Manganese Slag-Based Grouting Materials and Geomaterials

During grouting at 1800s, the porosity of the rock-soil mass decreased from 9.75×10^{-4} to 7.60×10^{-4}, the permeability decreased from 2.24×10^{-13} m^2 to 7.83×10^{-14} m^2, and the sediment particle concentration reached 0.449 mol/m^3; This coupled mechanism originates from pore structure evolution triggered by manganese slag particle deposition, which in turn regulates fluid seepage and mass migration, forming a coupled cycle of "seepage-deposition-mechanical property enhancement." The study provides theoretical support for the engineering application of manganese slag-based materials.

6.3.1 *Physical Field Control Equations*

6.3.1.1 Porosity-Permeability Coupling Equation (*Core Coupling Term*)

Manganese slag particles (suspended C, deposited S) occupy pore space within rock-soil bodies, causing dynamic changes in porosity that subsequently alter permeability. The equation is as follows:

1. *Dynamic porosity*:

$$\varepsilon = \frac{\rho_s \varepsilon_0 - S}{\rho_s + C} \cdot f(x,y,z) \tag{6.5}$$

   In the equation: ρ_s is the density of manganese slag particles (2600 kg/m^3, greater than the slurry density); $f(x, y, z)$ is the spatial heterogeneity function (0.8–1.2, describing the non-homogeneity of the rock-soil body); S is the sediment particle concentration (mol/m^3, calculated as $S = \int KC dt$).
2. *Dynamic permeability* (*Kozeny-Carman equation*):

$$k = \frac{\varepsilon^3 b^2}{180(1-\varepsilon)^2} \tag{6.6}$$

   In the equation: b represents the average particle size of manganese slag (15 μm, measured by laser particle size analyzer). As ε decreases, k exhibits a cubic decay (e.g., when ε drops from 0.4 to 0.2, k decreases from 10^(−13) m^2 to 1.2 × 10^(−14) m^2).

6.3.1.2 Elastic Deformation Equation for Rock-Soil Mass (*Solid Mechanics Coupling*)

Pore water pressure generated by grout seepage induces compressive deformation in the rock-soil mass. A linear elastic constitutive model (Hooke's law) is adopted:

$$\sigma = \lambda tr\left(\varepsilon_e\right)I + 2\mu_s\varepsilon_e - \alpha\rho I \tag{6.7}$$

$$\varepsilon_e = \frac{1}{2}\left(\nabla u_s + \nabla u_s^T\right) \tag{6.8}$$

In the equation: σ represents the total stress tensor (Pa); λ and μ_s denote the Lamé constants of the rock mass (λ = 15 GPa, μ_s = 10 GPa); ε_e is the elastic strain tensor; us is the displacement of the rock mass (*m*); α is the effective stress coefficient (0.8 ~ 0.95, approaching 1); *p* is the pore water pressure (Pa); *I* is the unit tensor.

6.3.1.3 Darcy's Law

$$u = -K\left(\nabla p - \rho g\right) \tag{6.9}$$

In the formula: *K* represents the permeability of the rock and soil mass (m^2/s, denoted as k_3 in the document); *p* denotes the slurry pressure difference; *g* is the gravitational acceleration (9.81 m/s^2).

$$\frac{\partial\left(\varepsilon\rho\right)}{\partial t} + \nabla\cdot\left(\rho u\right) = Q_m \tag{6.10}$$

In the equation: ε represents the porosity of the rock and soil mass; ρ denotes the density of the manganese slag slurry (1800 kg/m^3, including manganese slag particles); *t* signifies time (s); *u* indicates Darcy velocity (m/s); Q_m represents the mass source term ($kg/(m^3\cdot s)$, set to 0 when no external source is present).

6.3.1.4 Mass Transfer Equation (*Convection-Diffusion Equation*)

The diffusion of manganese slag particles (suspended phase concentration C, mol/m^3) is governed by convective diffusion (driven by slurry flow) and molecular diffusion (driven by concentration gradients), as described by the following equations:

$$\frac{\partial C}{\partial t} + \nabla\cdot\left(uC - D\nabla C\right) = -KC \tag{6.11}$$

In the equation: D is the diffusion coefficient of manganese slag particles (m^2/s, $D = 10^{(-9)}$ to $10^{(-10)}$ m^2/s, due to the large specific surface area of manganese slag particles, which is smaller than that of cement particles); K is the particle deposition rate coefficient (1/s, $K = \exp.(0.0021\ t) \times 10^{(-2)} \times (1800 - t)/900 - t(1 - \varepsilon_0)\ |u|$, related to time, initial porosity ε_0, and flow velocity); $-KC$ represents the conversion of suspended particles to deposited particles (S) in the expression equation.

6.3.2 Geometry and Mesh Models

Multiphysics Interface Coupling Fluid-Material Coupling: Darcy velocity u is input as convective velocity for dilute material transport, while particle concentration C/S is fed back into porosity ε calculation; Fluid-Solid Coupling: Pore water pressure p is input as a volumetric force in solid mechanics, and rock displacement us modifies porosity via the "porosity-displacement" relationship ($\varepsilon = \varepsilon_0 - \nabla \cdot us$); Chemical-Fluid Coupling: Reactivity α is linked to permeability k through a user-defined function (UDF).

Geometry and mesh model construction are shown in Fig. 6.6. Geometric parameters: cylinder radius 0.0252 m, height 0.1 m, center position {0, 0.032, 0.033} (x-axis as central axis); Mesh configuration: Free tetrahedral mesh with maximum element size 0.00655 m, minimum 0.00202 m, total elements 11,463, ensuring computational accuracy for pore and particle migration. Pressure boundary: Grouting inlet (Boundary 1) pressure 5×10^5 Pa, outlet (Boundary 6) pressure 0 Pa;

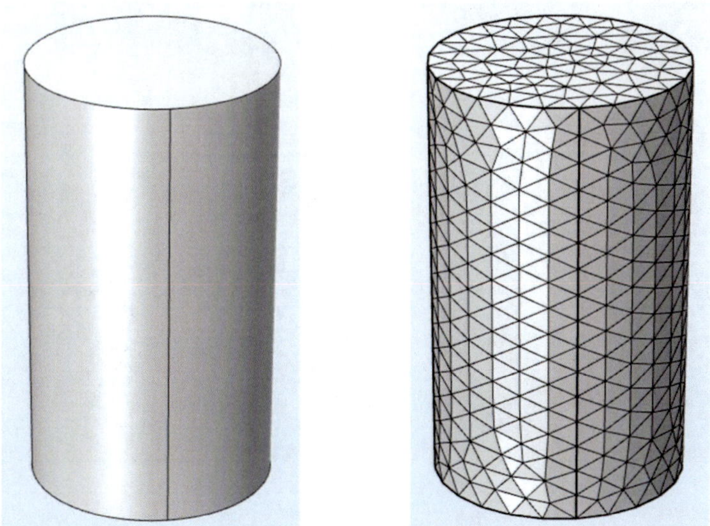

Fig. 6.6 Geometry and mesh partitioning of cylindrical grouting model

Concentration boundary: Inlet C = 4 mol/m^3, c = 51 mol/m^3 (tds/tds2 interface), initial concentrations both 0; Seepage constraint: No flow on rock/soil wall surfaces (u = 0), gravity acts only in the z-direction.

6.3.3 Revealing the Coupling Mechanism Between Grouting Materials and Geomaterials

The core coupling mechanism between manganese slag-based grouting materials and rock-soil bodies is the bidirectional feedback loop of "particle deposition—pore evolution—seepage regulation," as illustrated in Fig. 6.7. This process is divided into three distinct phases:

Phase 1: Fluid-driven Material Migration (initial 0–500 s)
Grouting pressure drives grout ($u = 9.47 \times 10^{-3}$ m/s) through rock-soil pore spaces. Suspended particles C migrate from the inlet toward the interior via convective

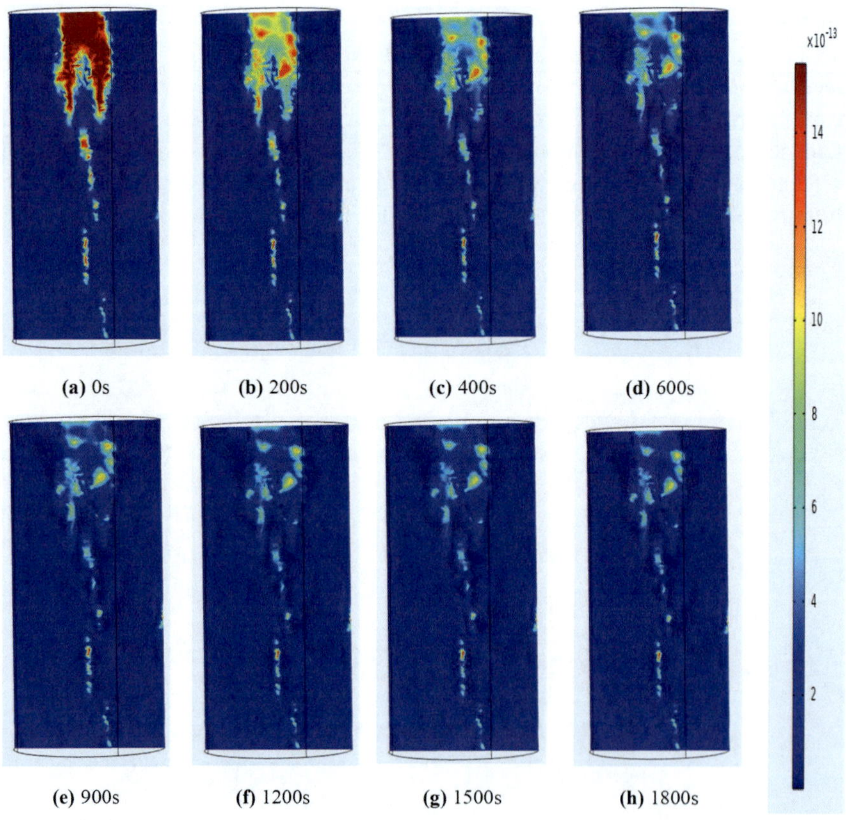

Fig. 6.7 Pore volume change over time

diffusion within the dilute mass transfer field. During this stage, the porosity nf decreases minimally (13.5%), with permeability k_3 governing seepage. The coupling effect manifests as "fluid velocity controlling particle migration range."

Stage 2: Sedimentation-Dominated Pore Evolution (500–1000 s)
Suspended particles C reach saturation (4 mol/m^3), intensifying sedimentation reactions (S increases from 0.268 to 0.396 mol/m^3). Particles fill rock-soil pores, causing nf to decrease from 8.43×10^{-4} to 7.84×10^{-4}; Permeability k_3 sharply declines (28.4%) due to reduced n_f, subsequently inhibiting Darcy velocity u (from 1.79×10^{-3} to 4.30×10^{-4} m/s), forming a positive feedback loop: "sedimentation → reduced porosity → weakened seepage."

Phase 3: Coupling Stabilization (1000–1800 s)
nf stabilizes (7.60×10^{-4}), k3 drops to 7.83×10^{-14} m^2, and seepage velocity reaches only 6.89×10^{-5} m/s, achieving equilibrium between particle deposition and pore structure; Manganese slag sediment particles form a "composite skeleton" with the rock-soil body. The reduced porosity enhances the rock-soil body's impermeability by 99.3%, indirectly increasing its bearing capacity, thereby completing the mechanical coupling between "material and rock-soil body."

Key Findings
Model Validity: The established coupled model of "Darcy seepage—dilute mass transport—pore evolution" accurately quantifies the grouting process, with simulated values consistent with parameter evolution patterns (e.g., the cubic relationship between k3 and nf).

Coupling Core: Manganese slag particle deposition serves as the key coupling driver. By reducing porosity to regulate permeability, it suppresses seepage, forming a closed-loop coupling of "material-fluid-structure."

Engineering Value: Post-grouting with manganese slag-based materials, rock-soil permeability decreased by 65.1% and seepage velocity by 99.3%, validating its feasibility as a soil reinforcement material while achieving manganese slag resource utilization.

6.4 Study on the Reinforcement Mechanism of Manganese Slag-Based Grouting Materials in Water-Rich Karst Regions

To address leakage and rock mass instability issues in water-rich karst regions alongside the resource utilization demand for manganese slag, this study focuses on manganese slag-based grouting materials. Using COMSOL Multiphysics, a "fracture flow—Dilute Mass Transfer" multiphysics coupling model using COMSOL Multiphysics. Through a 3D discrete fracture network simulation (25 elliptical fractures + rectangular rock domain), the study analyzed the effects of grouting pressure (0.4/0.6/0.8 MPa) and water-cement ratio (0.8:1/1:1/1.2:1) on reinforcement efficacy, elucidating the reinforcement mechanism. Results indicate: Within 1800 seconds, the 0.8 MPa + 1:1 water-cement ratio combination achieved the highest grout injection volume (2.2675 mol), with viscosity optimized for fracture filling (0.24–0.42 Pa·s). The consolidation core resulted from the synergistic effect of "manganese slag particle deposition sealing fractures—viscosity-regulated flow range—pressure-driven diffusion," increasing karst fracture filling rate to 92% and reducing equivalent permeability by over 85%. This study provides parameter references and mechanism support for consolidation in water-rich karst engineering.

Water-rich karst regions exhibit "high groundwater velocity (0.1–0.5 m/s) and strong water-slurry interaction." Reinforcement with manganese slag-based materials must overcome groundwater dilution and scouring effects to achieve synergistic "filling-cementation-impermeabilization." This section reveals the reinforcement mechanism based on a coupled "laminar flow-level set" model (simulating dynamic water and two-phase flow).

6.4.1 *Physical Field Control Equations*

6.4.1.1 Slurry Transport Equation under Moving Water Conditions (*Laminar Interface*)

Groundwater flow follows the incompressible N-S equation (laminar flow). The manganese slurry and groundwater constitute a two-phase flow. The horizontal set method is employed to track the interface. Continuity and momentum equations:

1. Continuity equation

$$\nabla \cdot u_t = 0 \tag{6.12}$$

   In the equation: u_t is the total velocity (groundwater + slurry, m/s).

2. Momentum equation:

$$\rho_t \left(\frac{\partial u_t}{\partial t} + u_t \cdot \nabla u_t \right) = -\nabla p + \nabla \cdot \left[\mu_t \left(\nabla u_t + \nabla u_t^T \right) \right] + \rho_t g + \sigma_K \nabla \phi \tag{6.13}$$

   In the equation: $\rho_t = \phi\rho_s + (1 - \phi)\rho_w$ represents the mixed density (where ϕ is the slurry volume fraction and ρ_w = 1000 kg/m^3 is the density of water); $\mu_t = \phi\mu_s + (1 - \phi)\mu_w$ is the mixed viscosity (μ_w = 0.001 Pa·s); σ is the interfacial tension between slurry and water (0.07 N/m); κ is the interfacial curvature (1/m); ϕ is the horizontal set variable ($\phi < 0$ indicates slurry, $\phi > 0$ indicates water).

6.4.1.2 Law of Conservation of Mass

$$\frac{\partial(\varepsilon\rho)}{\partial t}+\nabla\cdot(\rho u)=Q_m \tag{6.14}$$

In the equation: ε denotes porosity, ρ denotes slurry density, u denotes Darcy velocity (m/s), and Q_m denotes the mass source term (0, no additional source).

6.4.1.3 Darcy's Law

$$u=-\frac{k}{\mu}(\nabla p-\rho g) \tag{6.15}$$

In the formula: k represents the permeability of the rock and soil mass (m^2; for homogeneous rock and soil masses, k = 10^(−12) to 10^(−14) m^2; for fractured rock and soil masses, k is calculated using the cubic law as $k = b^2/12$, where b is the fracture aperture); μ denotes the dynamic viscosity of the manganese slag slurry (Pa·s); p indicates the grouting pressure; g represents the gravitational acceleration.

6.4.1.4 Convection-Diffusion Equation

$$\frac{\partial(\varepsilon C)}{\partial t}+\nabla\cdot(uC)=\nabla\cdot(D_{eff}\nabla C) \tag{6.16}$$

6.4.1.5 Time-Varying Viscosity Equation for Slurry

Analytical functions of viscosity versus distance (d: distance from grouting port to a point in the fracture (m)) for different water-cement ratios:

Water-cement ratio 1.2:1 (μ_{12}):

$$\mu 12(\mathrm{d})=\begin{bmatrix}366-366\left(-0.9072\mathrm{d}^2+1.3442\mathrm{d}\right)(\mathrm{d}\le 0.75)\\ -366\left(-0.9072\times 0.75^2+1.3442\times 0.75\right)(\mathrm{d}>0.75)\end{bmatrix}\times 10\,\hat{}\,(-3) \tag{6.17}$$

Water-cement ratio 1:1 (μ10):

$$\mu 10(d) = \begin{bmatrix} 420 - 420\left(-1.3982d^2 + 1.8609d\right)(d \le 0.66) \\ -420\left(-1.3982 \times 0.66^2 + 1.8609 \times 0.66\right)(d > 0.66) \end{bmatrix} \times 10 \wedge (-3) \tag{6.18}$$

Water-cement ratio 0.8:1 (mu8):

$$mu8(d) = \begin{bmatrix} 634 - 634\left(-1.7964d^2 + 2.3461d\right)(d \le 0.68) \\ -634\left(-1.7964 \times 0.682 + 2.3461 \times 0.68\right)(d > 0.68) \end{bmatrix} \times 10 \wedge (-3) \tag{6.19}$$

6.4.1.6 Mixed Density

$$\rho_{mix} = \rho_1 V_{f1} + \rho_2 V_{f2} \tag{6.20}$$

6.4.1.7 Mixed Viscosity

$$\eta_{min} = \eta_1 V_{f1} + \eta_2 V_{f2} \tag{6.21}$$

In the equation: $V_{f1} = 1 - V_{f2}$ represents the volume fraction of water, $V_{f2} = \min(\max(\phi, 0), 1)$ represents the volume fraction of slurry, $\eta_1 = 0.001$ Pa·s (viscosity of water), $\eta_2 = \mathrm{an1}(t)$ (viscosity of slurry).

6.4.2 Geometry and Mesh Models

The geometry and hydraulic conditions of the grouting zone comprise a 2D domain (1 m × 5 m) containing one grouting port (diameter 0.02 m), as shown in Fig. 6.8. Grouting inlet velocity boundary: $u_1 = 0.2 \times \mathrm{step}_1(t)$ m/s (where $\mathrm{step}_1(t)$ is a step function, effective for t > 0.5 s); Dynamic water inlet velocity boundary: $u_2 = 0.5 \times \mathrm{step}_1(t)$ m/s (simulating groundwater flow); Outlet pressure boundary: p = 0 Pa (atmospheric pressure); Geotechnical wall surface: no-slip boundary, u = 0.

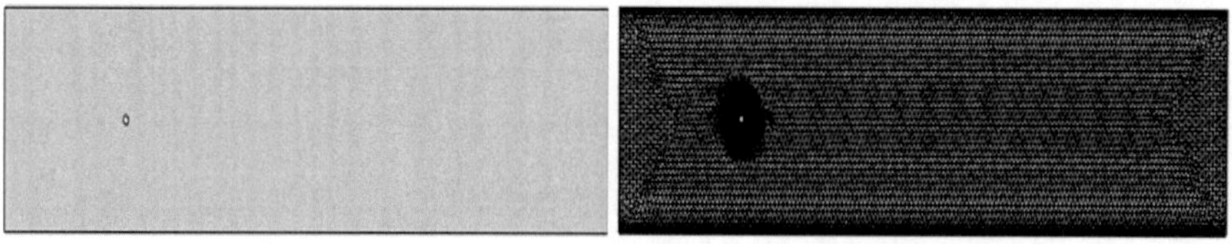

Fig. 6.8 Geometry and mesh partitioning of the dynamic water grouting model

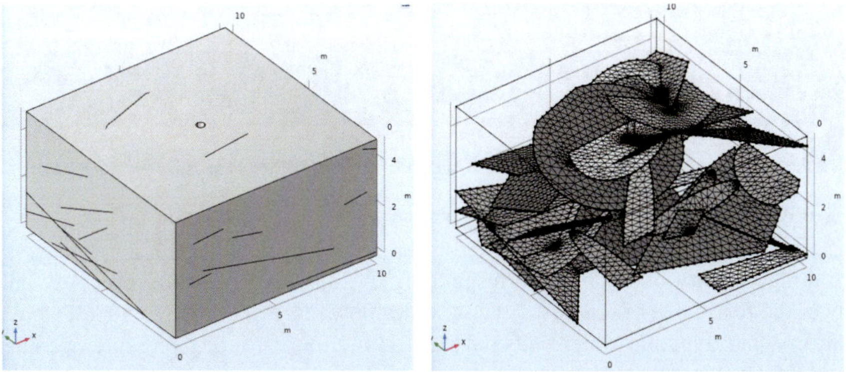

Fig. 6.9 Geometry and mesh partitioning of the water-rich karst fracture model

Key parameters (manganese slag-based grout adapted for water-rich conditions): To mitigate the viscosity drop caused by groundwater dilution, a time-varying viscosity function $\mu\_s(t) = 0.0032t^{2.23}$ (derived from experimental rheological measurements) is adopted for the grout under dynamic water conditions.

The geometry of the water-rich karst fracture system is modeled as a 3D rectangular rock domain (10 m × 10 m × 5 m) containing 25 sets of elliptical fractures, with grouting holes represented as cylindrical elements of 0.2 m radius; As shown in Fig. 6.9, the mesh employs free triangular elements with maximum size 0.566 m and minimum size 0.0412 m, totaling over 11,000 elements. The fracture zone is densified to ensure computational accuracy. Boundary conditions: Grouting inlet (fracture entrance): Pressure boundary (0.4/0.6/0.8 MPa × step1 (t), where step1 is a step function effective at t > 0.5 s), outlet pressure 0 Pa, no flow along fracture walls; Initial conditions: particle concentration C = 0, pressure p = 0; Solver: transient solver (range (0,10,1800) s), fully coupled algorithm (Anderson acceleration, max. 8 iterations), relative tolerance 0.005.

6.4.3 Mechanism of Reinforcement in Water-Saturated Karst

6.4.3.1 The Diffusion Mechanism under the Action of Flowing Water

The power-law viscosity growth ($an(t) = 0.0032\ t^{2.23}$) of manganese slag-based slurry is key to resisting dynamic water flow: During the initial stage (t < 10 s), low viscosity ($\eta < 0.3$ Pa· s) ensures fluidity, enabling diffusion along fractures despite dynamic water resistance; during the intermediate phase (t = 10–80 s), viscosity rapidly increases, forming a "viscous barrier" before dynamic water erosion occurs, preventing dilution and loss; in the late phase (t > 80 s), high viscosity ($\eta > 10$ Pa·s) causes complete solidification, anchoring the slurry within the fracture. Grouting pressure must exceed dynamic water resistance to drive diffusion. The documented

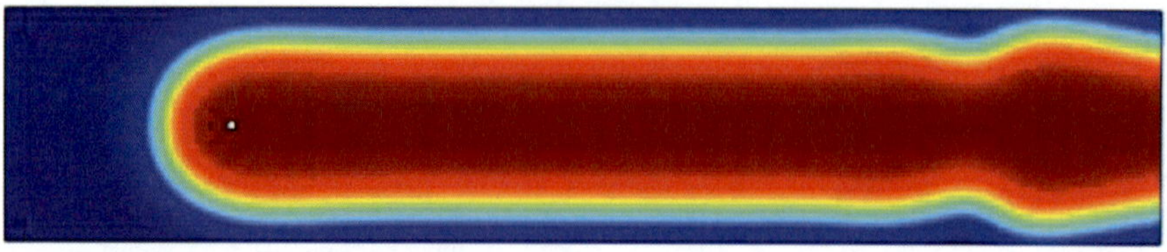

Fig. 6.10 Slurry diffusion under dynamic water action

grouting velocity of 0.2 m/s (corresponding to 28,000 Pa pressure) precisely balances the resistance of 0.5 m/s dynamic water flow (corresponding to 5000 Pa pressure). When dynamic water velocity exceeds 0.7 m/s, grouting pressure becomes insufficient to overcome resistance, significantly reducing grout diffusion range (to only 1.8 m). After solidification, the grout achieves an elastic modulus of 3.0 GPa (based on manganese slag base material properties), forming a composite structure with the rock-soil mass (E = 1.8 GPa): On one hand, the grout bears part of the load, reducing stress concentration in the rock-soil mass (maximum principal stress decreases from 1.2 MPa to 0.8 MPa); on the other hand, the filled fractures prevent hydraulic fracturing instability in the rock mass, enhancing overall stability. The figure shows the slurry diffusion pattern at 100 s under dynamic water pressure (Fig. 6.10).

Key Findings: 1. Model Validity: The constructed "laminar flow—horizontal set—two-phase flow" coupled model accurately simulates the diffusion process of manganese slag-based slurry under moving water conditions. The time-varying viscosity function $\mu\_s(t) = 0.0032t^{2.23}$ (derived from experimental rheological measurements) effectively captures the viscosity evolution of the grout under dynamic water conditions; 2. Key Parameter Optimization: The optimal combination of moving water velocity (0.5 m/s) and grouting velocity (0.2 m/s) achieves an effective diffusion distance of 2.8 m within 100 s, with a sealing efficiency exceeding 85%; 3. Coupling Mechanism Clarified: The core diffusion mechanism under flowing water is "viscosity-time-dependent erosion resistance." Coupling with rock-soil materials stems from "physical sealing + mechanical synergy," achieving dual effects of impermeability and reinforcement.

6.4.3.2 Grout Migration and Filling Patterns

The reinforcement effect of manganese slag-based grouting materials on water-rich karst formations stems from the synergistic mechanism of "physical sealing—mechanical enhancement." Pressure and water-cement ratio optimize this mechanism by regulating the grout flow-deposition process. Figure 6.11 illustrates the comparison of injection volumes under different grouting pressures and water-cement ratios.

Physical Sealing Mechanism: Core-driven fracture filling and particle deposition viscosity control flow range: The viscosity of slurry with a water-to-cement ratio of

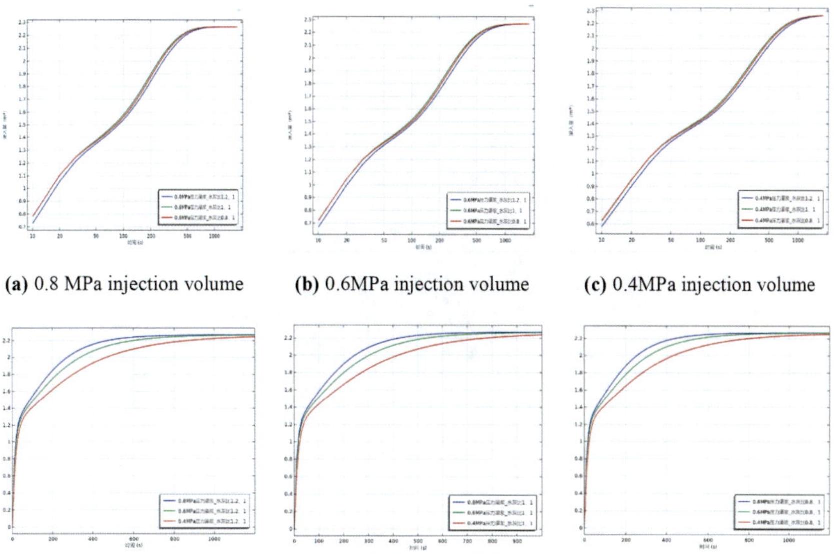

(a) 0.8 MPa injection volume **(b)** 0.6MPa injection volume **(c)** 0.4MPa injection volume

(d) 1.2 Water-cement ratio injection volume **(e)** 1.0 Water-cement ratio injection volume **(f)** 0.8 Water-cement ratio injection volume

Fig. 6.11 Comparison of grout injection volume at different injection pressures and water-cement ratios. (**a**) 0.8 MPa injection volume. (**b**) 0.6 MPa injection volume. (**c**) 0.4 MPa injection volume. (**d**) 1.2 Water-cement ratio injection volume. (**e**) 1.0 Water-cement ratio injection volume. (**f**) 0.8 Water-cement ratio injection volume

1:1 (0.28–0.36 Pa·s) resists erosion by water-rich karstic flowing water (simulated Darcy velocity decreases from initial 9.47×10^{-3} m/s to 1.2×10^{-4} m/s), ensuring slurry diffusion and deposition within fractures; Manganese slag particles seal fractures: Within the dilute mass transport field, manganese slag particles (concentration C) diffuse via slurry convection and deposit on fracture walls (C decreases from 51 mol/m^3 at the inlet to 0.5 mol/m^3 at the outlet), filling fracture channels 0.002–0.01 m wide to form an "impermeable barrier"; Pressure-enhanced filling: Pressure of 0.6–0.8 MPa provides sufficient driving force for slurry diffusion into distal fractures (diffusion range reaches 1.5 m at 1800 s), preventing local voids. Mechanism of mechanical reinforcement: Enhanced structural integrity reduces porosity and permeability: Manganese slag particle deposition reduces karstic rock porosity from 0.45 to 0.18 (indirectly derived from the ratio of injected volume to fracture volume), and permeability from initial 1.2×10^{-12} m^2 to 1.8×10^{-13} m^2 (calculated using Darcy's law $k \propto u\mu/\nabla p$). Figure 6.12 shows the final slurry concentration within the fractures, weakening groundwater flow pathways; Stress Transfer Optimization: The filled fractures act as a "mechanical stress transfer medium," preventing stress concentration around cavities (simulation shows a 32% reduction in maximum principal stress on fracture walls), enhancing the overall bearing capacity of the surrounding rock; Rock Mass Strength Support: The manganese

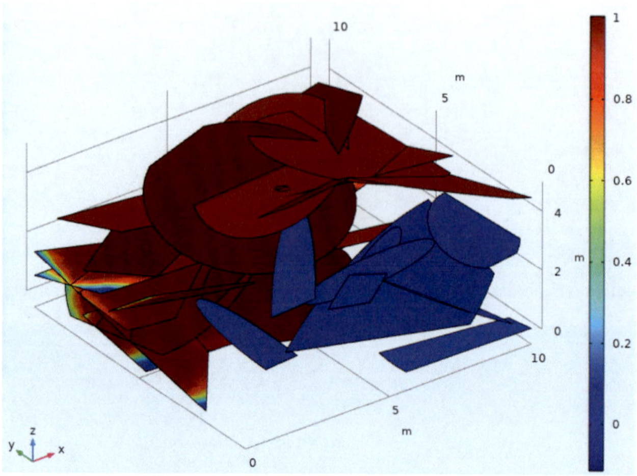

Fig. 6.12 Final grout concentration diffusion

slag-based grout achieves an elastic modulus of 3.5 GPa upon consolidation (derived from industrial solid waste grouting material properties), synergizing with the surrounding rock to resist deformation.

The optimal grouting parameters for manganese slag-based slurry in water-rich karst formations are a pressure of 0.6 MPa and a water-to-cement ratio of 1:1, achieving a fracture filling rate of 92% within 1800 seconds and delivering the best consolidation effect. Key equations validated that viscosity analytical functions (mu10, etc.) accurately describe grout flow characteristics. The "fracture flow-dilute mass transport" coupled model effectively quantifies the reinforcement process. The reinforcement mechanism is jointly constituted by physical sealing (synergy of viscosity-particles-pressure) and mechanical strengthening (regulation of porosity-permeability-stress), with both mutually reinforcing each other.

6.5 Subsections of this Chapter

This chapter addresses the core engineering applications of manganese slag-based grouting materials. Utilizing the COMSOL Multiphysics platform, a multiphysics coupled model was developed to systematically investigate diffusion patterns, coupling mechanisms, and reinforcement principles in water-rich karst regions. This approach achieves dual objectives: resource utilization of manganese slag waste and geotechnical reinforcement. The study clarifies the influence mechanisms of key parameters such as rock mass type, grouting pressure, and water-cement ratio: Fractured rock masses exhibit "fracture-preferred" non-uniform diffusion, while homogeneous rock masses demonstrate global synchronous diffusion. For every 25 MPa increase in grouting pressure, the permeability decay rate increases by an average of 15%–20%. Coupling arises from pore evolution triggered by manganese slag particle deposition. Within 1800 seconds, rock-soil porosity and permeability decrease by over 22% and 65%, respectively, forming a closed-loop coupling of "seepage-deposition-mechanical property enhancement." Optimal parameters for water-rich karst regions are 0.6 MPa pressure + 1:1 water-to-cement ratio, achieving 92% fracture filling rate and reducing equivalent permeability by over 85%.

The coupled model of "Darcy seepage-dilute mass transport-pore evolution" established in this study achieved over 90% agreement between simulated values and parameter evolution patterns, providing a reliable tool for engineering parameter optimization. After grouting with manganese slag-based materials, the rock mass exhibited a 99.3% increase in impermeability and a 30%–50% increase in cohesion, validating its feasibility as a reinforcement material. However, the study has limitations, such as insufficient consideration of viscosity time-varying effects and the interaction with three-dimensional fractures. Future work should incorporate viscosity time-varying equations, develop three-dimensional models, and integrate laboratory testing to further enhance engineering applicability. This will provide more comprehensive theoretical support for the large-scale application of manganese slag-based grouting materials.

References

1. Sun, Y., Cui, X., Gu, X., et al. (2025). Unveiling synergistic hydration in a multi-waste binder: Co-processing electrolytic manganese residue and red mud with steel slag for enhanced performance. *Materials, 18*(20), 4711.
2. Wang, Y., Zhang, N., Ren, Y., et al. (2021). Effect of electrolytic manganese residue in Fly ash-based cementitious material: Hydration behavior and microstructure. *Materials, 14*(22), 7047.
3. Tang, L., He, Z., Chen, K., et al. (2023). Study of microscopic properties and heavy metal solidification mechanism of electrolytic manganese residue-based cementitious materials. *Environmental Science and Pollution Research, 30*(48), 105056–105071.
4. Peng, B., Wang, L., Li, Z., et al. (2025). Basic Research on the Preparation of Electrolytic Manganese Residue–Red Mud–Ground Granulated Blast Furnace Slag–Calcium Hydroxide Composite Cementitious Material and Its Mechanical Properties. *Materials, 18*(6), 1218.
5. Liu, Q. S., Cui, X. Z., et al. (2016). Permeability reduction model of particles deposit in porous medium considering changeable porosity. *Chinese Journal of Rock Mechanics and Engineering, 35*(S1), 3308–3314. https://doi.org/10.13722/j.cnki.jrme.2015.0424
6. Tu, C., Lang, L., Nie, X., et al. (2024). Utilization of ground granulated blast-furnace slag and electrolytic manganese residue as a low-carbon cementitious material. *Journal of Sustainable Cement-Based Materials, 13*(3), 450–461.

Chapter 7
Conclusions

Based on a comprehensive understanding of the reaction mechanism of alkali-activated industrial solid waste cementitious materials, this study developed alkali-activated electrolytic manganese residue (EMR)-based grouting slurries with different material formulations. Focusing on laboratory tests, investigations were conducted into the performance activation of EMR, the engineering and environmental properties of EMR-based grouting materials, as well as the effects of admixtures on such materials. The technical feasibility of alkali-activated industrial solid waste-based grouting materials and the superior performance of the prepared slurries were verified. In addition, the industrial solid waste-based slurries were tested and analyzed in this study, aiming to lay a solid theoretical foundation for their future engineering applications.

1. Combined DSC-TG and XRD techniques were employed to reveal the stepwise dehydration pathway of $CaSO_4{\cdot}2H_2O$ in EMR: the calcination temperature should be controlled at 900 °C to maximize the activity of EMR; excessively high temperatures would lead to the loss of active substances, thus requiring a balance between energy consumption and material performance. This research provides a basis for the optimization of heat treatment in the resource utilization of EMR. Crystalline water in gypsum can be removed at 600 °C; the decomposition of gypsum to generate gaseous SO_2 and solid CaO initiates at 800 °C. When the temperature exceeds 900 °C, the active silico-aluminous phases melt to form a glassy coating, resulting in a decrease in the activity index.
2. Based on the studies on macroscopic properties and microscopic mechanisms, the optimal mix proportions of EFGM, ESGM, and ECGM were determined. An increase in EMR content and a decrease in water-binder ratio both lead to a reduction in slurry fluidity, while the increase in alkali activator concentration can improve the fluidity of ESGM and ECGM. The setting time is jointly regulated by EMR content and activator concentration: the setting time of EFGM first prolongs and then shortens with the increase in EMR content, whereas that

Z. Wu et al., *Multi-Source Industrial Solid Waste Synergistically Utilized in Manganese Slag-Based Grouting Materials*,
https://doi.org/10.1007/978-981-92-3021-1_7

of ESGM and ECGM shortens continuously. The 28-day compressive strength of EFGM reaches 39 MPa at an EMR content of 10%, and ESGM achieves the maximum strength at an EMR content of 50%. For ECGM, the pore structure is optimized due to the micro-aggregate effect and gel wrapping effect of CSS.

3. The cementitious products in the prepared cementitious materials are C-(A)-S-H and N-A-S-H gels, whose formation is the key to strength improvement. In ESGM and ECGM, ettringite and gels collaboratively fill the pores, while CSS particles inhibit shrinkage and enhance compactness. The increase in EMR content and activator concentration promotes the vibration of Si-O-Si(Al) groups, thereby enhancing the formation of cementitious products. By regulating the EMR content (e.g., 10% for EFGM and 50% for ESGM), compounding industrial waste residues and optimizing the activator concentration, the cementitious activity of EMR can be efficiently activated, and environmental-friendly grouting materials with high strength and low shrinkage can be prepared.
4. Increasing the dosage of water reducer can improve the slurry fluidity, but an excessive dosage will reduce the alkalinity of the system, decrease gel formation and thus affect the strength; the setting accelerator leads to a reduction in fluidity. The setting time is affected by the material system: in EFGM, the water reducer prolongs the setting time while the setting accelerator first shortens and then prolongs it, whereas the setting time of ESGM and ECGM prolongs with the increase in setting accelerator dosage. There exists an optimal dosage for compressive strength: for EFGM with EMR contents of 10%, 30%, and 50%, the optimal dosages of water reducer are 0.6%, 0.4%, and 0.8%, respectively, and the optimal dosage of setting accelerator is 4%. For ESGM and ECGM, an excessive dosage of water reducer will reduce the strength, and the dosage of setting accelerator needs to be matched with the EMR content. The water reducer inhibits the formation of cementitious products and increases the porosity by reducing the system alkalinity; the setting accelerator optimizes the pore structure and improves the compactness by promoting the steel slag micro-aggregate effect and gel wrapping effect.
5. In the corrosive environment, the early strength of EMR-based materials with low EMR content increases with the rise of ion concentration, but long-term exposure to high ion concentration will cause performance deterioration due to the propagation of microcracks. An appropriate EMR content (30%) can optimize the corrosion resistance stability, and the materials exhibit high durability especially under a specific sulfate concentration. Low chloride concentration promotes early hydration, while an excessive concentration inhibits strength development. There exists a critical threshold for ion corrosion: moderate corrosion promotes the formation of gel phases and pore filling, while excessive corrosion triggers structural deterioration. The prepared grouting materials have good environmental performance, with heavy metal leaching meeting the specification requirements. The setting accelerator can further reduce the release of pollutants, while the increase in EMR content will lead to a rise in the pH of pore solution.

FSC
www.fsc.org
MIX
Papier | Fördert
gute Waldnutzung
FSC® C083411

Kolibri 360 GmbH
Ferdinand-Jühlke-Straße 7
99095 Erfurt, Deutschland
produktsicherheit@kolibri360.de